*Solutions Manual to Accompany*
# Introduction to Time Series Analysis and Forecasting

T0214939

WILEY SERIES IN PROBABILITY AND STATISTICS

Established by WALTER A. SHEWART and SAMUEL S. WILKS

Editors: *David J. Balding, Noel A. C. Cressie, Nicholas I. Fisher,
Iain M. Johnstone, J. B. Kadane, Geert Molenberghs, Louise M. Ryan,
David W. Scott, Adrian F. M. Smith, Jozef L. Teugels*
Editors Emeriti: *Vic Barnett, J. Stuart Hunter, David G. Kendall*

*Solutions Manual to Accompany*
# Introduction to Time Series Analysis and Forecasting

DOUGLAS C. MONTGOMERY
CHERYL L. JENNINGS
MURAT KULAHCI

*Prepared by*

JAMES R. BROYLES
RACHEL T. JOHNSON
CHRISTOPHER J. RIGDON
*Arizona State Universtiy*

WILEY
A JOHN WILEY & SONS, INC., PUBLICATION

Copyright © 2009 by John Wiley & Sons, Inc. All rights reserved.

Published by John Wiley & Sons, Inc., Hoboken, New Jersey.
Published simultaneously in Canada.

No part of this publication may be reproduced, stored in a retrieval system, or transmitted in any form or by any means, electronic, mechanical, photocopying, recording, scanning, or otherwise, except as permitted under Section 107 or 108 of the 1976 United States Copyright Act, without either the prior written permission of the Publisher, or authorization through payment of the appropriate per-copy fee to the Copyright Clearance Center, Inc., 222 Rosewood Drive, Danvers, MA 01923, 978-750-8400, fax 978-646-8600, or on the web at www.copyright.com. Requests to the Publisher for permission should be addressed to the Permissions Department, John Wiley & Sons, Inc., 111 River Street, Hoboken, NJ 07030,
(201) 748-6011, fax (201) 748-6008, or online at http://www.wiley.com/go/permission.

Limit of Liability/Disclaimer of Warranty: While the publisher and author have used their best efforts in preparing this book, they make no representations or warranties with respect to the accuracy or completeness of the contents of this book and specifically disclaim any implied warranties of merchantability or fitness for a particular purpose. No warranty may be created or extended by sales representatives or written sales materials. The advice and strategies contained herein may not be suitable for your situation. You should consult with a professional where appropriate. Neither the publisher nor author shall be liable for any loss of profit or any other commercial damages, including but not limited to special, incidental, consequential, or other damages.

For general information on our other products and services or for technical support, please contact our Customer Care Department within the U.S. at 877-762-2974, outside the U.S. at 317-572-3993 or fax 317-572-4002.

Wiley also publishes its books in a variety of electronic formats. Some content that appears in print, however, may not be available in electronic format. For more information about Wiley products, visit our web site at www.wiley.com.

*Library of Congress Cataloging in Publication Data:*

Montgomery, Douglas C.
 Solutions Manual to Accompany Introduction to Time Series Analysis and Forecasting

 ISBN 978-0-470-43574-8

10 9 8 7 6 5 4 3 2

# ➢ <u>Contents</u>

# Contents

# ➤ Preface

The purpose of this solutions manual is to provide the student with an understanding of how to apply the concepts presented in *Introduction to Time Series Analysis and Forecasting* by Douglas C. Montgomery, Cheryl L. Jennings, and Murat Kulahci.

This manual contains exercise solutions to primarily the odd numbered problems. Not all presented solutions are complete, and some solutions are omitted to reduce repetition.

The exercises were solved using the following statistical software packages widely available for the personal computer:

➤ JMP® Version 7, www.jmp.com

➤ MINITAB™ Version 14, www.minitab.com

# Chapter 1

## INTRODUCTION TO FORECASTING

## Exercises

**1.1** In any organization or business, accurate prediction of future events, such as sales, stock prices, yield, virus outbreaks, or weather patterns, is critical for long-term success. Statistical forecasting methods present an objective set of tools to forecast these future events based on past events (data).

**1.2** A time series is a time-oriented or chronological sequence of observations on a variable of interest. A trend effect is one that has a noticeable increasing or decreasing movement. Seasonal variation is the tendency of a time series to vary in a cyclic pattern, such as monthly or quarterly. Random error is noise present in the system. The noise can fluctuate between measurements.

**1.3** A point forecast does not take into account the inherent variability in the forecast. A prediction interval is a range of values for the future observation.

**1.4** Causal forecasting techniques involve the forecast of one variable based on how it is affected by changes in another variable.

**1.6** A rolling horizon forecast system updates or revises the forecast for all of the previous time periods in the horizon and computes a forecast for the newest time period.

**1.7** The forecast horizon is the number of future periods for which forecasts must be produced. The horizon is often dictated by the nature of the problem. The forecast interval is the frequency with which new forecasts are prepared.

# Chapter 2

## STATISTICS BACKGROUND FOR FORECASTING

## Exercises

### 2.3

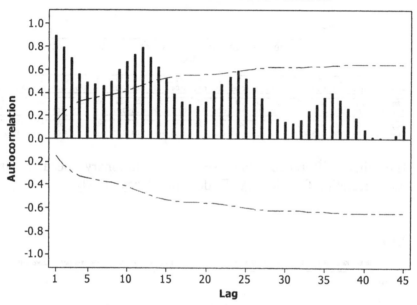

**Beer Sales Autocorrelation Function**

| Lag | 1 | 2 | 3 | 4 | 5 | 6 | 7 | 8 | 9 | 10 | 11 | 12 | 13 | 14 | 15 | 16 | 17 | 18 | 19 | 20 | 21 | 22 | 23 |
|---|---|---|---|---|---|---|---|---|---|---|---|---|---|---|---|---|---|---|---|---|---|---|---|
| ACF | 0.89 | 0.80 | 0.70 | 0.56 | 0.49 | 0.48 | 0.46 | 0.50 | 0.60 | 0.67 | 0.73 | 0.79 | 0.71 | 0.62 | 0.52 | 0.39 | 0.32 | 0.30 | 0.28 | 0.32 | 0.41 | 0.47 | 0.54 |

| Lag | 24 | 25 | 26 | 27 | 28 | 29 | 30 | 31 | 32 | 33 | 34 | 35 | 36 | 37 | 38 | 39 | 40 | 41 | 42 | 43 | 44 | 45 |
|---|---|---|---|---|---|---|---|---|---|---|---|---|---|---|---|---|---|---|---|---|---|---|
| ACF | 0.59 | 0.52 | 0.44 | 0.35 | 0.23 | 0.17 | 0.15 | 0.13 | 0.16 | 0.24 | 0.30 | 0.36 | 0.40 | 0.33 | 0.26 | 0.18 | 0.08 | 0.02 | 0.01 | 0.00 | 0.03 | 0.12 |

The time series is nonstationary because the autocorrelation function (ACF) is persistent (decays very slowly). Seasonality is also present.

**2.7**

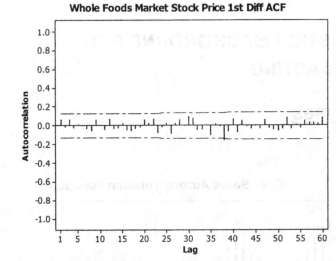

**Whole Foods Market Stock Price 1st Diff ACF**

| Lag | 1 | 2 | 3 | 4 | 5 | 6 | 7 | 8 | 9 | 10 | 11 | 12 | 13 | 14 | 15 | 16 | 17 | 18 | 19 | 20 |
|-----|---|---|---|---|---|---|---|---|---|----|----|----|----|----|----|----|----|----|----|----|
| ACF | 0.07 | -0.02 | 0.07 | -0.03 | 0.01 | 0.00 | -0.04 | -0.06 | 0.07 | 0.00 | -0.04 | 0.07 | -0.04 | -0.03 | 0.03 | -0.05 | -0.05 | -0.04 | -0.02 | 0.07 |

| Lag | 21 | 22 | 23 | 24 | 25 | 26 | 27 | 28 | 29 | 30 | 31 | 32 | 33 | 34 | 35 | 36 | 37 | 38 | 39 | 40 |
|-----|----|----|----|----|----|----|----|----|----|----|----|----|----|----|----|----|----|----|----|----|
| ACF | 0.03 | 0.07 | -0.08 | -0.02 | 0.03 | -0.09 | 0.03 | 0.06 | 0.00 | 0.09 | 0.08 | -0.05 | -0.05 | 0.01 | -0.10 | 0.02 | -0.01 | -0.16 | -0.07 | 0.07 |

| Lag | 41 | 42 | 43 | 44 | 45 | 46 | 47 | 48 | 49 | 50 | 51 | 52 | 53 | 54 | 55 | 56 | 57 | 58 | 59 | 60 |
|-----|----|----|----|----|----|----|----|----|----|----|----|----|----|----|----|----|----|----|----|----|
| ACF | -0.07 | 0.06 | -0.01 | -0.03 | -0.01 | -0.03 | 0.07 | -0.03 | -0.04 | -0.06 | -0.03 | 0.09 | -0.04 | 0.02 | -0.02 | 0.06 | 0.04 | 0.03 | 0.03 | 0.09 |

The first difference time series is stationary because the autocorrelation function (ACF) does not decay slowly.

**2.13**

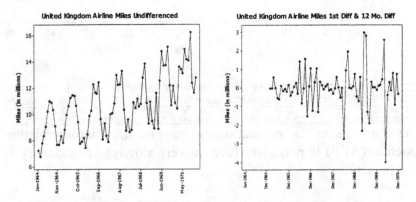

The first differencing removes the trend from the original data.

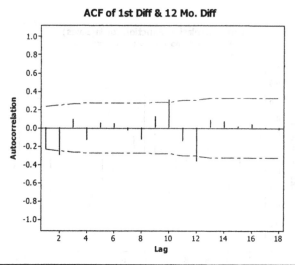

ACF of 1st Diff & 12 Mo. Diff

| Lag | 1 | 2 | 3 | 4 | 5 | 6 | 7 | 8 | 9 | 10 | 11 | 12 | 13 | 14 | 15 | 16 | 17 | 18 |
|-----|-----|-----|-----|-----|-----|-----|-----|-----|-----|-----|-----|-----|-----|-----|-----|-----|-----|-----|
| ACF | -0.23 | -0.30 | 0.10 | -0.14 | 0.06 | 0.05 | -0.02 | -0.12 | 0.12 | 0.31 | -0.15 | -0.36 | 0.09 | 0.07 | 0.02 | 0.04 | 0.00 | -0.02 |

The first difference ACF's rapid decay suggests that the time series is stationary.

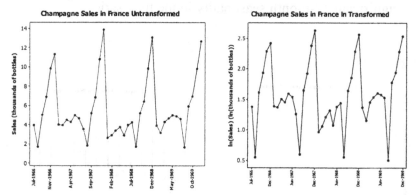

**a.** The log transformation made the time series more symmetric (less skewed) about the overall average. The transformed data are closer to the normal distribution than the skewed untransformed data.

5

**b.**

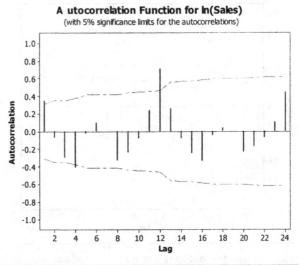

**A utocorrelation Function for ln(Sales)**
(with 5% significance limits for the autocorrelations)

| Lag | 1 | 2 | 3 | 4 | 5 | 6 | 7 | 8 | 9 | 10 | 11 | 12 | 13 | 14 | 15 | 16 | 17 | 18 | 19 | 20 | 21 | 22 | 23 | 24 |
|---|---|---|---|---|---|---|---|---|---|---|---|---|---|---|---|---|---|---|---|---|---|---|---|---|
| ACF | 0.35 | -0.07 | -0.29 | -0.40 | -0.02 | 0.10 | 0.00 | -0.33 | -0.24 | -0.08 | 0.23 | 0.71 | 0.26 | -0.07 | -0.25 | -0.34 | -0.04 | 0.05 | 0.01 | -0.23 | -0.16 | -0.07 | 0.11 | 0.45 |

**c.** The sample autocorrelation function implies that there is significant 12-month seasonality in the time series.

**2.21**

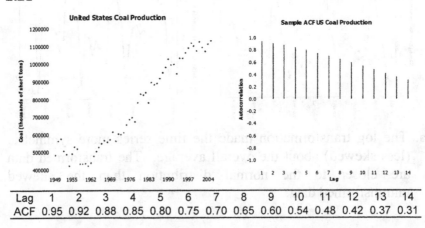

| Lag | 1 | 2 | 3 | 4 | 5 | 6 | 7 | 8 | 9 | 10 | 11 | 12 | 13 | 14 |
|---|---|---|---|---|---|---|---|---|---|---|---|---|---|---|
| ACF | 0.95 | 0.92 | 0.88 | 0.85 | 0.80 | 0.75 | 0.70 | 0.65 | 0.60 | 0.54 | 0.48 | 0.42 | 0.37 | 0.31 |

The time series is nonstationary. This can be seen from the time series plot or the very slow decay of the autocorrelation function.

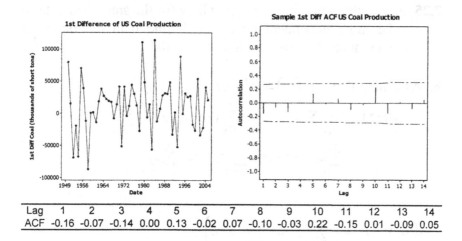

| Lag | 1 | 2 | 3 | 4 | 5 | 6 | 7 | 8 | 9 | 10 | 11 | 12 | 13 | 14 |
|-----|---|---|---|---|---|---|---|---|---|----|----|----|----|----|
| ACF | -0.16 | -0.07 | -0.14 | 0.00 | 0.13 | -0.02 | 0.07 | -0.10 | -0.03 | 0.22 | -0.15 | 0.01 | -0.09 | 0.05 |

First differencing has made the time series stationary.

**2.23**

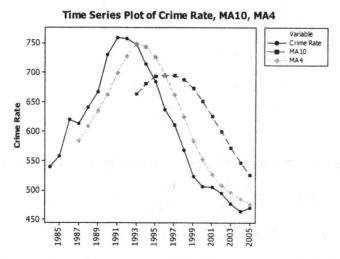

The 10-period moving average (MA10) has smoothed the data drastically but considerably lags the original crime rate. The 4-period moving average (MA4) has also smoothed the data but lags the original crime rate much less.

**2.25  a.** Weighted moving averages allow for the smoothed data to be more dependent on selected time than others; more weight can be given to more recent data.

**b.** Assuming that $\mathrm{Var}(y_t) = \sigma^2 \ \forall t$,

$$\mathrm{Var}(M_T^w) = \mathrm{Var}\left(\sum_{t=T-N+1}^{T} a_{T+1-t}\, y_t\right) = \sum_{t=T-N+1}^{T} \mathrm{Var}(a_{T+1-t}\, y_t) = \sum_{t=T-N+1}^{T} a_{T+1-t}^2\, \mathrm{Var}(y_t)$$

$$= \sum_{t=T-N+1}^{T} a_{T+1-t}^2 \sigma^2 = \sigma^2 \sum_{t=T-N+1}^{T} a_{T+1-t}^2 = \sigma^2 \sum_{j=1}^{N} a_j^2$$

**c.**

$$\mathrm{Cov}(M_T^w, M_{T+k}^w) = \mathrm{Cov}\left(\sum_{t=T-N+1}^{T} a_{T+1-t}\, y_t,\ \sum_{v=T+k-N+1}^{T+k} a_{T+k+1-v}\, y_v\right)$$

$$= E\left[\left(\sum_{t=T-N+1}^{T} a_{T+1-t}\, y_t - E\left[\sum_{t=T-N+1}^{T} a_{T+1-t}\, y_t\right]\right)\left(\sum_{v=T+k-N+1}^{T+k} a_{T+k+1-v}\, y_v - E\left[\sum_{v=T+k-N+1}^{T+k} a_{T+k+1-v}\, y_v\right]\right)\right]$$

$$= E\left[\left(\sum_{t=T-N+1}^{T} a_{T+1-t}\,(y_t - \mu)\right)\left(\sum_{v=T+k-N+1}^{T+k} a_{T+k+1-v}\,(y_v - \mu)\right)\right]$$

$$= E\left[\sum_{t=T-N+1}^{T}\sum_{v=T+k-N+1}^{T+k} a_{T+1-t}\, a_{T+k+1-v}\,(y_t - \mu)(y_v - \mu)\right]$$

$$= \sum_{t=T-N+1}^{T}\sum_{v=T+k-N+1}^{T+k} a_{T+1-t}\, a_{T+k+1-v}\, E[(y_t - \mu)(y_v - \mu)]$$

$$= \sum_{t=T-N+1}^{T}\sum_{v=T+k-N+1}^{T+k} a_{T+1-t}\, a_{T+k+1-v}\, \mathrm{Cov}[(y_t - \mu)(y_v - \mu)]$$

$$= \sum_{t=T+k-N+1}^{T} a_{T+1-t}\, a_{T+k+1-t}\, \sigma^2 = \sigma^2 \sum_{j=1}^{N-k} a_j a_{j+k} \qquad \text{for } |k| < N$$

Because the time series data points are independent of each other,

$$\mathrm{Cov}[(y_t - \mu)(y_v - \mu)] = \begin{cases} \sigma^2 & \text{if } t = v \\ 0 & \text{if } t \neq v \end{cases}.$$

If $|k| \geq N$, then $\mathrm{Cov}(M_T^w, M_{T+k}^w) = 0$.

**d.**

$$\rho_k = \frac{\mathrm{Cov}(M_T^w, M_{T+k}^w)}{\mathrm{Var}(M_T^w, M_{T+k}^w)} = \begin{cases} \dfrac{\sigma^2 \sum_{j=1}^{N-k} a_j a_{j+k}}{\sigma^2 \sum_{j=1}^{N} a_j^2} = \dfrac{\sum_{j=1}^{N-k} a_j a_{j+k}}{\sum_{j=1}^{N} a_j^2} & \text{for } k = 1,\ldots,N-1 \\[4mm] 0 & \text{for } k \geq N \end{cases}$$

**2.27**

$$\mathrm{Var}(e_{T+1}(1)) = \mathrm{Var}(y_{T+1} - M_T) = \mathrm{Var}(y_{T+1}) + \mathrm{Var}(M_T) = \sigma^2 + \mathrm{Var}\left(\frac{\sum\limits_{i=T-N+1}^{T} y_i}{N}\right) = \sigma^2 + \frac{1}{N^2}\sum_{i=T-N+1}^{T}\mathrm{Var}(y_i)$$

$$= \sigma^2 + \frac{1}{N^2}(N\sigma^2) = \sigma^2 + \frac{\sigma^2}{N} = \sigma^2\left(\frac{N+1}{N}\right)$$

**2.29** For $T \le t_1 - 1$,

$$E[M_T] = E\left[\frac{\sum\limits_{t=T-N+1}^{T} y_t}{N}\right] = \frac{1}{N}E\left[\sum_{t=T-N+1}^{T} y_t\right] = \frac{1}{N}\sum_{t=T-N+1}^{T}E[y_t] = \frac{N\mu}{N} = \mu \cdot$$

For $t_1 \le T \le t_1 + N - 1$,

$$E[M_T] = E\left[\frac{\sum\limits_{t=T-N+1}^{T} y_t}{N}\right] = \frac{1}{N}E\left[\sum_{t=T-N+1}^{T} y_t\right] = \frac{1}{N}E\left[\sum_{t=T-N+1}^{t_1-1} y_t + \sum_{t=t_1}^{T} y_t\right] = \frac{1}{N}\left(E\left[\sum_{t=T-N+1}^{t_1-1} y_t\right] + E\left[\sum_{t=t_1}^{T} y_t\right]\right)$$

$$= \frac{1}{N}\left(\sum_{t=T-N+1}^{t_1-1}E[y_t] + \sum_{t=t_1}^{T}E[y_t]\right) = \frac{1}{N}\left((t_1 - T + N - 1)\mu + (T - t_1 + 1)(\mu + \delta)\right)$$

$$= \frac{1}{N}(t_1\mu - T\mu + N\mu - \mu + T\mu - t_1\mu + \mu + T\delta - t_1\delta + \delta) = \frac{1}{N}(N\mu + (T - t_1 + 1)\delta)$$

$$= \mu + \frac{T - t_1 + 1}{N}\delta$$

For $T \ge t_1 + N$,

$$E[M_T] = E\left[\frac{\sum\limits_{t=T-N+1}^{T} y_t}{N}\right] = \frac{1}{N}E\left[\sum_{t=T-N+1}^{T} y_t\right] = \frac{1}{N}\sum_{t=T-N+1}^{T}E[y_t] = \frac{N(\mu + \delta)}{N} = \mu + \delta$$

**2.31**

$$E[M_T] = E\left[\frac{\sum_{t=T-N+1}^{T} y_t}{N}\right] = \frac{1}{N}E\left[\sum_{t=T-N+1}^{T} y_t\right] = \frac{1}{N}\sum_{t=T-N+1}^{T} E[y_t] = \frac{1}{N}\sum_{t=T-N+1}^{T}(\beta_0 + \beta_1 t) = \frac{1}{N}\left(N\beta_0 + \beta_1\sum_{t=T-N+1}^{T} t\right)$$

$$= \beta_0 + \frac{\beta_1}{N}\sum_{j=1}^{N}(T-N+j) = \beta_0 + \frac{\beta_1}{N}\left(NT - N^2 + \sum_{j=1}^{N} j\right) = \beta_0 + \frac{\beta_1}{N}\left(NT - N^2 + \frac{N(N+1)}{2}\right)$$

$$= \beta_0 + \beta_1 T - \beta_1 N + \frac{\beta_1(N+1)}{2} = \beta_0 + \beta_1 T - \left(\frac{2N-N-1}{2}\right)\beta_1 = \beta_0 + \beta_1 T - \frac{N-1}{2}\beta_1$$

**2.35**  **a.** The forecast error autocorrelation function indicates that there is significant autocorrelation at lag 4.

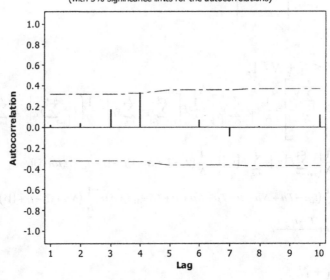

**Autocorrelation Function for et(1)**
(with 5% significance limits for the autocorrelations)

**b.** The normal probability plot indicates that the errors are normally distributed.

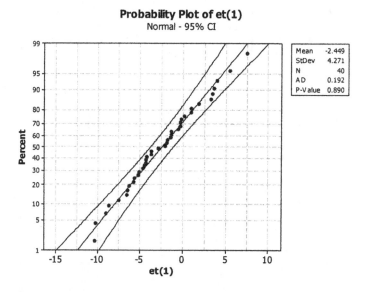

**Probability Plot of et(1)**
Normal - 95% CI

| | |
|---|---|
| Mean | -2.449 |
| StDev | 4.271 |
| N | 40 |
| AD | 0.192 |
| P-Value | 0.890 |

**d.** The mean error is −2.45, mean squared error is 23.78, and mean absolute error is 4.01. These results indicate that the forecast is bias because the mean error is significantly different than zero, the mean squared error is significantly larger than unity, and the mean absolute error is significantly large.

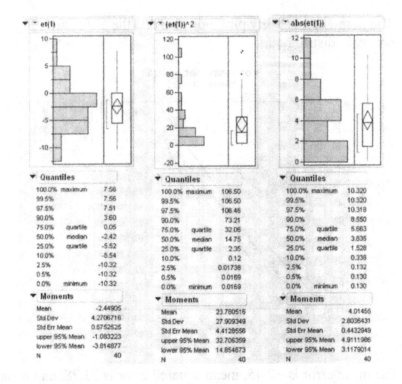

| ▼ et(1) Quantiles | | |
|---|---|---|
| 100.0% | maximum | 7.56 |
| 99.5% | | 7.56 |
| 97.5% | | 7.51 |
| 90.0% | | 3.60 |
| 75.0% | quartile | 0.05 |
| 50.0% | median | -2.42 |
| 25.0% | quartile | -5.52 |
| 10.0% | | -8.54 |
| 2.5% | | -10.32 |
| 0.5% | | -10.32 |
| 0.0% | minimum | -10.32 |

| ▼ et(1) Moments | |
|---|---|
| Mean | -2.44905 |
| Std Dev | 4.2706716 |
| Std Err Mean | 0.6752525 |
| upper 95% Mean | -1.083223 |
| lower 95% Mean | -3.814877 |
| N | 40 |

| ▼ (et(1))^2 Quantiles | | |
|---|---|---|
| 100.0% | maximum | 106.50 |
| 99.5% | | 106.50 |
| 97.5% | | 106.46 |
| 90.0% | | 73.21 |
| 75.0% | quartile | 32.06 |
| 50.0% | median | 14.75 |
| 25.0% | quartile | 2.35 |
| 10.0% | | 0.12 |
| 2.5% | | 0.01738 |
| 0.5% | | 0.0169 |
| 0.0% | minimum | 0.0169 |

| ▼ (et(1))^2 Moments | |
|---|---|
| Mean | 23.780516 |
| Std Dev | 27.909349 |
| Std Err Mean | 4.4128556 |
| upper 95% Mean | 32.706359 |
| lower 95% Mean | 14.854673 |
| N | 40 |

| ▼ abs(et(1)) Quantiles | | |
|---|---|---|
| 100.0% | maximum | 10.320 |
| 99.5% | | 10.320 |
| 97.5% | | 10.318 |
| 90.0% | | 8.550 |
| 75.0% | quartile | 5.663 |
| 50.0% | median | 3.835 |
| 25.0% | quartile | 1.528 |
| 10.0% | | 0.338 |
| 2.5% | | 0.132 |
| 0.5% | | 0.130 |
| 0.0% | minimum | 0.130 |

| ▼ abs(et(1)) Moments | |
|---|---|
| Mean | 4.01456 |
| Std Dev | 2.8036431 |
| Std Err Mean | 0.4432949 |
| upper 95% Mean | 4.9111966 |
| lower 95% Mean | 3.1179014 |
| N | 40 |

**2.39** The forecasting system does not seem to be stable because there is a significant decreasing trend in the errors ($p$-value = 0.009).

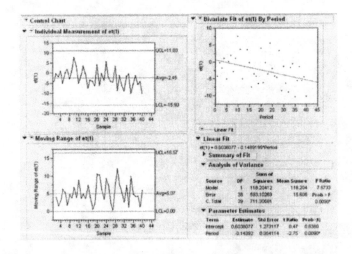

# Chapter 3

# REGRESSION ANALYSIS AND FORECASTING

## Exercises

**3.1 a.**

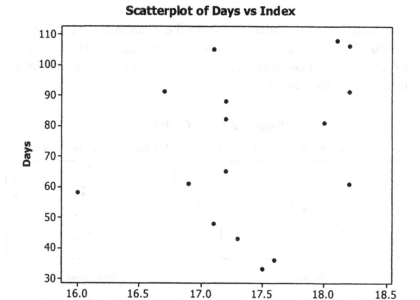

**b.** The regression equation is: $Days = -111 + 10.5\ Index$

Minitab Output

| Predictor | Coef | SE Coef | T | P |
|-----------|------|---------|------|-------|
| Constant | -110.6 | 180.6 | -0.61 | 0.550 |
| Index | 10.51 | 10.37 | 1.01 | 0.328 |

S = 25.0364   R-Sq = 6.8%   R-Sq(adj) = 0.2%
PRESS = 10809.2   R-Sq(pred) = 0.00%

**c.**

Minitab Output: Analysis of Variance

| Source | DF | SS | MS | F | P |
|--------|-----|--------|-------|------|-------|
| Regression | 1 | 644.0 | 644.0 | 1.03 | 0.328 |
| Residual Error | 14 | 8775.5 | 626.8 | | |
| Total | 15 | 9419.4 | | | |

**d.** 95% confidence and prediction intervals at Index = 17:

Minitab Output

| Fit | SE Fit | 95% CI | 95% PI |
|-------|--------|------------------|-------------------|
| 68.04 | 7.54 | (51.86, 84.22) | (11.96, 124.13) |

**e.** The model seems inadequate from the residuals. The residuals versus fitted values have a cyclical structure. The graph of residuals versus observation order indicates that autocorrelation is present in the residuals.

**Residual Plots for Days**

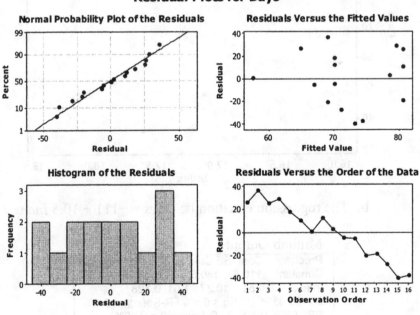

14

**f.** The Durbin-Watson statistic of 0.138086 indicates the autocorrelation is significantly different than zero. See Table A.5 for approximate critical values of the Durbin-Watson test statistic.

**3.5** The $n$ observations should be split evenly at the $-1$ and $1$ locations. This constructs the D-optimal design that minimizes the variance of the slope estimate.

**3.9**

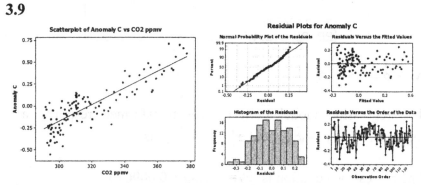

The regression equation is:
*Anomaly C = −3.03 + 0.00953 CO2 ppmv*

Minitab Output

| Predictor | Coef | SE Coef | T | P |
|---|---|---|---|---|
| Constant | -3.0303 | 0.1677 | -18.07 | 0.000 |
| CO2 ppmv | 0.0095311 | 0.0005273 | 18.07 | 0.000 |

S = 0.134748  R-Sq = 72.6%  R-Sq(adj) = 72.4%
PRESS = 2.30497  R-Sq(pred) = 71.77%

Analysis of Variance

| Source | DF | SS | MS | F | P |
|---|---|---|---|---|---|
| Regression | 1 | 5.9316 | 5.9316 | 326.68 | 0.000 |
| Residual Error | 123 | 2.2333 | 0.0182 | | |
| Total | 124 | 8.1649 | | | |

Durbin-Watson statistic = 1.22572

The residuals of the model fit look good but the Durbin-Watson statistic indicates that there is a significant positive autocorrelation function of the residuals. The first iteration of the Cochran-Orcutt method to estimate the parameters is as follows:

The regression model of the residual estimates is $\hat{\phi} = 0.38575$.
The regression equation is: $\varepsilon_t = 0.386\, \varepsilon_{t-1}$

Minitab Output

| Predictor | Coef | SE Coef | T | P |
|---|---|---|---|---|
| $\varepsilon_{t-1}$ | 0.38575 | 0.08366 | 4.61 | 0.000 |

S = 0.124392
PRESS = 1.92977

with $\hat{\phi} = 0.38575$.

The regression model built on the transformed response and regressor $y'_t = y_t - \hat{\phi}y_{t-1}$ and $x'_t = x_t - \hat{\phi}x_{t-1}$ is:

The regression equation is: $y'_t = -1.89 + 0.00967\, x'_t$

Minitab Output

| Predictor | Coef | SE Coef | T | P |
|---|---|---|---|---|
| Constant | -1.8892 | 0.1537 | -12.29 | 0.000 |
| xtprime | 0.0096681 | 0.0007848 | 12.32 | 0.000 |

S = 0.124284   R-Sq = 55.4%   R-Sq(adj) = 55.1%
PRESS = 1.94851   R-Sq(pred) = 53.92%

Therefore the parameter estimates are $\hat{\beta}'_0 = -1.8892$ and $\hat{\beta}_1 = 0.009681$.

**3.13** The model fit after the first iteration of the Cochrane-Orcutt procedure in Exercise 3.12 is

$$\hat{y}'_{t3.12} = \hat{\beta}'_{03.12} + \hat{\beta}_1 x'_{t3.12} = -0.67063 + 0.296499 x'_{t3.12}$$ with output:

16

Minitab Output

| Predictor | Coef | SE Coef | T | P |
|---|---|---|---|---|
| Constant | -0.67063 | 0.03265 | -20.54 | 0.000 |
| xt_prime3.12 | 0.296499 | 0.009373 | 31.63 | 0.000 |

S = 0.0137887   R-Sq = 98.5%   R-Sq(adj) = 98.4%
PRESS = 0.00359171   R-Sq(pred) = 98.14%

The model fit for the first differences $y'_t = y_t - y_{t-1}$ and $x'_t = x_t - x_{t-1}$ through the origin is $\hat{y}'_t = \hat{\beta}_1 x'_t = 0.29220 x'_{t3.12}$ with output:

Minitab Output

| Predictor | Coef | SE Coef | T | P |
|---|---|---|---|---|
| xt_prime | 0.29220 | 0.02818 | 10.37 | 0.000 |

S = 0.0160532                    PRESS = 0.00471464

Both estimates of the slope $\beta_1$ are very similar.

**3.17** Recall that the error is $e_i = y_i - \hat{y}_i$.

$$SS_R = SS_T - SS_E = \sum_{i=1}^{n}(y_i - \bar{y})^2 - \sum_{i=1}^{n}(y_i - \hat{y}_i)^2 = \sum_{i=1}^{n}(y_i^2 - 2y_i\bar{y} + \bar{y}^2) - \sum_{i=1}^{n}(y_i^2 - 2y_i\hat{y}_i + \hat{y}_i^2)$$

$$= \sum_{i=1}^{n}y_i^2 - \sum_{i=1}^{n}y_i^2 - \sum_{i=1}^{n}2y_i\bar{y} + \sum_{i=1}^{n}\bar{y}^2 + \sum_{i=1}^{n}2y_i\hat{y}_i - \sum_{i=1}^{n}\hat{y}_i^2$$

$$= -2\bar{y}\sum_{i=1}^{n}y_i + \sum_{i=1}^{n}\bar{y}^2 + \sum_{i=1}^{n}2y_i\hat{y}_i - \sum_{i=1}^{n}\hat{y}_i^2$$

$$= -2n\bar{y}^2 + n\bar{y}^2 + \sum_{i=1}^{n}2y_i\hat{y}_i - \sum_{i=1}^{n}\hat{y}_i^2 = -n\bar{y}^2 + 2\sum_{i=1}^{n}\hat{y}_i(\hat{y}_i + e_i) - \sum_{i=1}^{n}\hat{y}_i^2$$

$$= -n\bar{y}^2 + 2\sum_{i=1}^{n}\hat{y}_i^2 + \sum_{i=1}^{n}e_i - \sum_{i=1}^{n}\hat{y}_i^2 = -n\bar{y}^2 + 2\sum_{i=1}^{n}\hat{y}_i^2 + 0 - \sum_{i=1}^{n}\hat{y}_i^2 = \sum_{i=1}^{n}\hat{y}_i^2 - n\bar{y}^2$$

A property of least squares regression is that $\sum_{i=1}^{n}e_i = 0$.

**3.21** The simple linear regression model is $y = \beta_0 + \beta_1 t + \varepsilon$.

$$\text{Cov}(\hat{\underline{\beta}}) = \sigma^2 \underline{C} = \sigma^2(\underline{X}'\underline{X})^{-1} = \sigma^2 \left( \begin{bmatrix} 1 & 1 & \cdots & 1 \\ 1 & 2 & \cdots & T \end{bmatrix} \begin{bmatrix} 1 & 1 \\ 1 & 2 \\ \vdots & \vdots \\ 1 & T \end{bmatrix} \right)^{-1} = \sigma^2 \begin{bmatrix} T & \sum\limits_{t=1}^{T} t \\ \sum\limits_{t=1}^{T} t & \sum\limits_{t=1}^{T} t^2 \end{bmatrix}^{-1}$$

$$= \sigma^2 \begin{bmatrix} T & \dfrac{T(T+1)}{2} \\ \dfrac{T(T+1)}{2} & \dfrac{T(T+1)(2T+1)}{6} \end{bmatrix}^{-1} = \sigma^2 \left( \dfrac{12}{T^2(T^2-1)} \right) \begin{bmatrix} \dfrac{T(T+1)(2T+1)}{6} & -\dfrac{T(T+1)}{2} \\ -\dfrac{T(T+1)}{2} & T \end{bmatrix}$$

$$= \sigma^2 \begin{bmatrix} \dfrac{2(2T-1)}{T(T-1)} & -\dfrac{6}{T(T-1)} \\ -\dfrac{6}{T(T-1)} & \dfrac{12}{T(T^2-1)} \end{bmatrix}$$

$$V(\hat{\beta}_0) = \sigma^2 C_{00} = \sigma^2 \dfrac{2(2T-1)}{T(T-1)} \ , \qquad V(\hat{\beta}_1) = \sigma^2 C_{11} = \sigma^2 \dfrac{12}{T(T^2-1)}$$

**3.23** There is no indication that any point is a high leverage point because all the $h_{ii}$ values ($h$ usage) are less than $2p/n = 2(2)/12 = 1/3 \approx 0.333$. There is no indication of influential points because all the Cook's $D$ values are less than 1.

### JMP Output

**Response Usage**

**Whole Model**

**Summary of Fit**

| | |
|---|---|
| RSquare | 0.999865 |
| RSquare Adj | 0.999852 |
| Root Mean Square Error | 1.945628 |
| Mean of Response | 421.8617 |
| Observations (or Sum Wgts) | 12 |

**Analysis of Variance**

| Source | DF | Sum of Squares | Mean Square | F Ratio |
|---|---|---|---|---|
| Model | 1 | 280589.57 | 280590 | 74122.78 |
| Error | 10 | 37.85 | 3.78547 | Prob > F |
| C. Total | 11 | 280627.42 | | <.0001* |

**Lack Of Fit**

| Source | DF | Sum of Squares | Mean Square | F Ratio |
|---|---|---|---|---|
| Lack Of Fit | 9 | 36.323448 | 4.03594 | 2.8357 |
| Pure Error | 1 | 1.531250 | 1.53125 | Prob > F |
| Total Error | 10 | 37.854698 | | 0.4469 |
| | | | | Max RSq |
| | | | | 1.0000 |

**Parameter Estimates**

| Term | Estimate | Std Error | t Ratio | Prob>|t| |
|---|---|---|---|---|
| Intercept | -6.332087 | 1.870046 | -3.79 | 0.0035* |
| Temp | 9.2084678 | 0.033823 | 272.25 | <.0001* |

| | Month | Temp | Usage | Cook's D Influence Usage | h Usage |
|---|---|---|---|---|---|
| 1 | Jan | 21 | 185.79 | 0.11238479 | 0.27984285 |
| 2 | Feb | 24 | 214.47 | 0.00216543 | 0.23632517 |
| 3 | Mar | 32 | 288.03 | 0.00254302 | 0.14687217 |
| 4 | Apr | 47 | 424.84 | 0.03486584 | 0.08340688 |
| 5 | May | 50 | 454.68 | 0.00477991 | 0.08703536 |
| 6 | Jun | 59 | 539.03 | 0.097036 | 0.13055304 |
| 7 | Jul | 68 | 621.55 | 0.14206837 | 0.22302811 |
| 8 | Aug | 74 | 675.06 | 0.00010374 | 0.3118767 |
| 9 | Sep | 62 | 562.03 | 0.18999965 | 0.15593835 |
| 10 | Oct | 50 | 452.93 | 0.01860073 | 0.08703536 |
| 11 | Nov | 41 | 369.95 | 0.02373576 | 0.09247507 |
| 12 | Dec | 30 | 273.98 | 0.51740426 | 0.16560895 |

**3.27** Using mixed stepwise regression with both the probability of entering and the probability of leaving equal to 0.25:

$$\hat{y} = 78.8 + 42.2x_1 + 10.2x_2 + 26.6x_3 + 5.2x_4 + 9.9x_1x_2 + 14.0x_1x_3 + 8.6x_1x_4$$

JMP Output

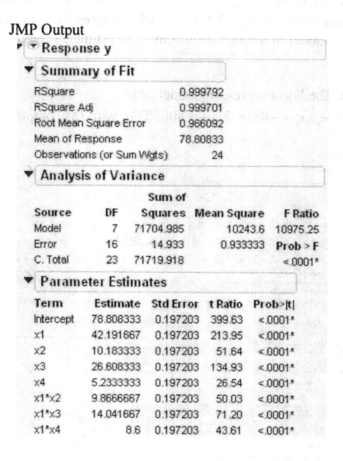

▶ ▼ Response y

▼ Summary of Fit

| | |
|---|---|
| RSquare | 0.999792 |
| RSquare Adj | 0.999701 |
| Root Mean Square Error | 0.966092 |
| Mean of Response | 78.80833 |
| Observations (or Sum Wgts) | 24 |

▼ Analysis of Variance

| Source | DF | Sum of Squares | Mean Square | F Ratio |
|---|---|---|---|---|
| Model | 7 | 71704.985 | 10243.6 | 10975.25 |
| Error | 16 | 14.933 | 0.933333 | Prob > F |
| C. Total | 23 | 71719.918 | | <.0001* |

▼ Parameter Estimates

| Term | Estimate | Std Error | t Ratio | Prob>|t| |
|---|---|---|---|---|
| Intercept | 78.808333 | 0.197203 | 399.63 | <.0001* |
| x1 | 42.191667 | 0.197203 | 213.95 | <.0001* |
| x2 | 10.183333 | 0.197203 | 51.64 | <.0001* |
| x3 | 26.608333 | 0.197203 | 134.93 | <.0001* |
| x4 | 5.2333333 | 0.197203 | 26.54 | <.0001* |
| x1*x2 | 9.8666667 | 0.197203 | 50.03 | <.0001* |
| x1*x3 | 14.041667 | 0.197203 | 71.20 | <.0001* |
| x1*x4 | 8.6 | 0.197203 | 43.61 | <.0001* |

**3.29** Fitting the simple linear model yields
$$\hat{y}_t = \hat{\beta}_0 + \hat{\beta}_1 x_t = -1.158 + 0.2924 x_t, \text{ with error slope}$$
$$\hat{\phi} = 0.434.$$
Therefore the fitted model for the autocorrelated errors Eq. (3.119) is
$$\hat{y}_t = \hat{\phi} y_{t-1} + \hat{\beta}_0 + \hat{\beta}_1 x_t + \hat{\beta}_2 x_{t-1} = 0.434 y_{t-1} - 0.671 + 0.266 x_t - 0.098 x_{t-1}.$$

The fitted model for the autocorrelated errors Eq. (3.120) is
$$\hat{y}_t = \phi y_{t-1} + \hat{\beta}_0 + \hat{\beta}_1 x_t = 0.434 y_{t-1} - 0.666 + 0.170 x_t.$$

The model from Eq. (3.119) has a $\hat{\beta}_1$ that is close to the one estimated using the Eq. (3.120) model. In Exercise 3.13, the first iteration model fit through the origin slope estimate is large.

**3.35 a.** The linear regression model fit is
$$\hat{y}_i = \hat{\beta}_0 + \hat{\beta}_1 x_i = -857432.3 + 3.8005073 x_i \text{ with JMP output:}$$

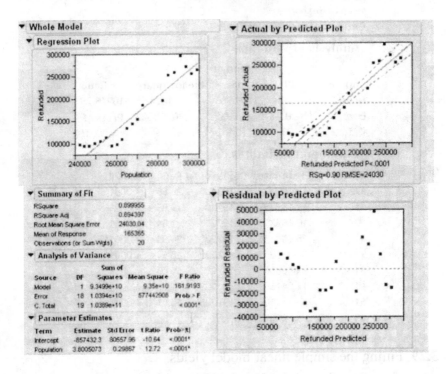

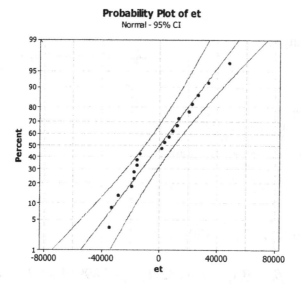

**Probability Plot of et**
Normal - 95% CI

There appears to be a nonlinear effect because of the U-shaped residual versus refunded plot. The residuals appear normally distributed.

There appears to be significant lag 1 correlation in the residuals with $\hat{\phi} = 0.5888276$:

JMP Output

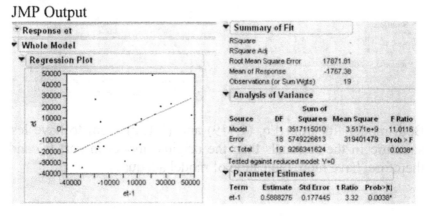

**b.** The model fit of Eq. (3.119) is

$$\hat{y}_t = \hat{\phi} y_{t-1} + \hat{\beta}_0 + \hat{\beta}_1 x_t + \hat{\beta}_2 x_{t-1} = 0.5888276 y_{t-1} - 392767 + 0.63037 x_t + 1.10959 x_{t-1}.$$

JMP Output

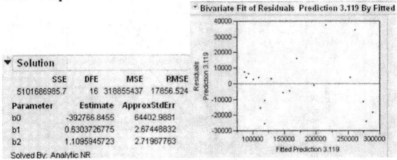

**Solution**

| | SSE | DFE | MSE | RMSE |
|---|---|---|---|---|
| | 5101686985.7 | 16 | 318855437 | 17856.524 |

| Parameter | Estimate | ApproxStdErr |
|---|---|---|
| b0 | -392766.8455 | 64402.9881 |
| b1 | 0.6303726775 | 2.67448832 |
| b2 | 1.1095945723 | 2.71967763 |

Solved By: Analytic NR

The model fit of Eq. (3.120) is

$$\hat{y}_t = \phi y_{t-1} + \hat{\beta}_0 + \hat{\beta}_1 x_t = 0.5888276 y_{t-1} - 389944 + 1.71727 x_t.$$

JMP Output

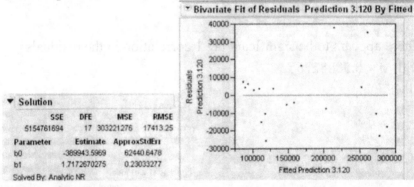

**Solution**

| | SSE | DFE | MSE | RMSE |
|---|---|---|---|---|
| | 5154761694 | 17 | 303221276 | 17413.25 |

| Parameter | Estimate | ApproxStdErr |
|---|---|---|
| b0 | -389943.5969 | 62440.6478 |
| b1 | 1.7172670275 | 0.23033277 |

Solved By: Analytic NR

Both model fits from Eq. (1.119) and (1.120) seem to have less structure in the residuals than the original model in part (a) and, therefore, have more indication of model adequacy.

# Chapter 4

# Exponential Smoothing Methods

## Exercises

**4.1**   **a.**

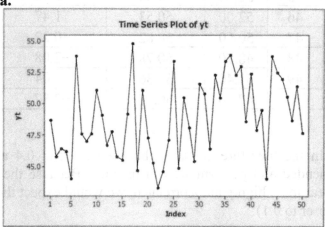

**b.** The smoothing procedure works well. The method captures the general trend in the data and there is a considerable reduction in the variability of the data.

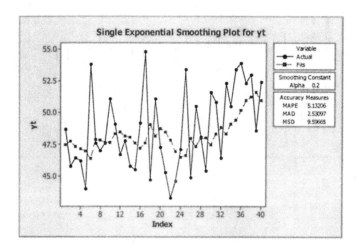

**c.**

| Period | Actual | Forecast | Forecast Error |
|--------|--------|----------|----------------|
| 41 | 47.90 | 51.26 | −3.36 |
| 42 | 49.50 | 50.59 | −1.09 |
| 43 | 44.00 | 50.37 | −6.37 |
| 44 | 53.80 | 49.10 | 4.70 |
| 45 | 52.50 | 50.04 | 2.46 |
| 46 | 52.00 | 50.53 | 1.47 |
| 47 | 50.60 | 50.82 | −0.22 |
| 48 | 48.70 | 50.78 | −2.08 |
| 49 | 51.40 | 50.36 | 1.04 |
| 50 | 47.70 | 50.57 | −2.87 |

**4.3**     In the literature, $\lambda$ values between 0.1 and 0.4 are often recommended and perform well in practice. Because the data do not appear to be highly autocorrelated, we would expect the $\lambda$ to be low (closer to 0.1).

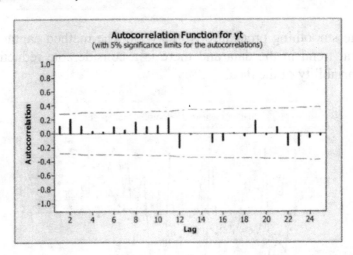

**4.5**    **a.** The Minitab Output for the single exponential smoothing fit is shown below.

Minitab Output

> **Single Exponential Smoothing for E.42 Data**
> \* ERROR \* Illegal weight - must be between 0 and 2.

The data indicate that there is an error in finding the optimum value for the exponential smoothing constant $\lambda$. In this case, the data are very noisy and a value of $\lambda = 0$ is in the confidence interval range for the optimum $\lambda$.

**b.** Because the optimum $\lambda$ was approximately zero, the one-step-ahead forecasts for the last 10 observations will be equal to the mean.

**4.7**    **a.**

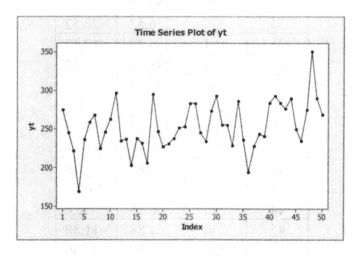

**b.** The smoothing procedure works well; it captures the trend of the data and reduces the variance.

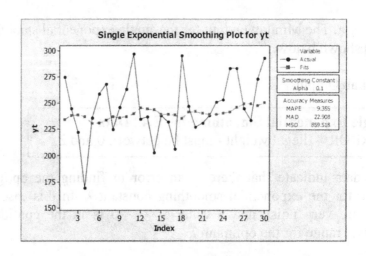

**c.**

| Period | Actual | Forecast | Error |
|--------|--------|----------|-------|
| 31 | 255 | 256.16 | −1.16 |
| 32 | 255 | 256.05 | −1.05 |
| 33 | 229 | 255.94 | −26.94 |
| 34 | 286 | 253.25 | 32.75 |
| 35 | 236 | 256.52 | −20.52 |
| 36 | 194 | 254.47 | −60.47 |
| 37 | 228 | 248.42 | −20.42 |
| 38 | 244 | 246.38 | −2.38 |
| 39 | 241 | 246.14 | −5.14 |
| 40 | 284 | 245.63 | 38.37 |
| 41 | 293 | 249.47 | 43.53 |
| 42 | 284 | 253.82 | 30.18 |
| 43 | 276 | 256.84 | 19.16 |
| 44 | 290 | 258.75 | 31.25 |
| 45 | 250 | 261.88 | −11.88 |
| 46 | 235 | 260.69 | −25.69 |
| 47 | 275 | 258.12 | 16.88 |
| 48 | 350 | 259.81 | 90.19 |
| 49 | 290 | 268.83 | 21.17 |
| 50 | 269 | 270.95 | −1.95 |

26

**d.** The forecast error falls near the 3σ limit only once (σ estimated by the moving range chart). We could conclude that the forecasting procedure works well; however, the time series is very noisy as are the forecast errors.

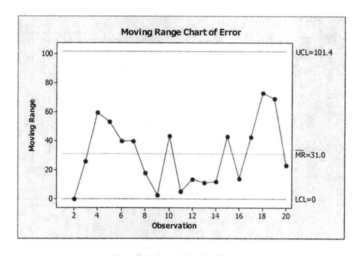

$$\hat{\sigma} = \frac{\overline{MR}}{1.128} = \frac{31.0}{1.128} = 27.5$$

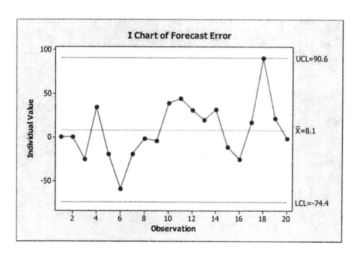

**4.9**    Yes, differencing has removed the linear upward trend in the data.

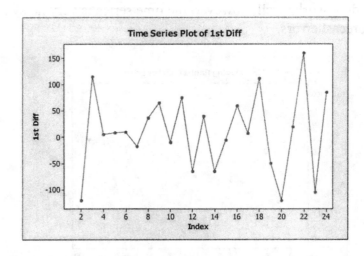

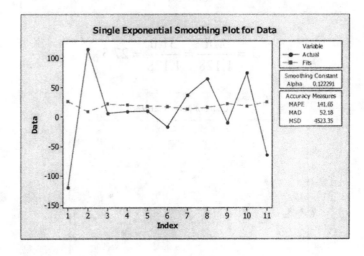

| Period | First Difference | Difference Forecasts | Original Data | Forecast for Original Data |
|--------|------------------|---------------------|---------------|---------------------------|
| 13 | 40 | 14.10 | 460 | 474.10 |
| 14 | −65 | 17.27 | 395 | 412.27 |
| 15 | −5 | 7.22 | 390 | 397.22 |
| 16 | 60 | 5.72 | 450 | 455.72 |
| 17 | 8 | 12.36 | 458 | 470.36 |
| 18 | 112 | 11.82 | 570 | 581.82 |
| 19 | −50 | 24.06 | 520 | 544.06 |
| 20 | −120 | 15.01 | 400 | 415.01 |
| 21 | 20 | −1.48 | 420 | 418.52 |
| 22 | 160 | 1.14 | 580 | 581.14 |
| 23 | −105 | 20.55 | 475 | 495.55 |
| 24 | 85 | 5.21 | 560 | 565.21 |

**4.13    a.** The optimum value of $\lambda$ as given by Minitab is 0.0005443. This produces a mean squared deviation (MSD) of 47402.2, which is lower than the MSD of 50533.0 produced by using a value of $\lambda = 0.1$.

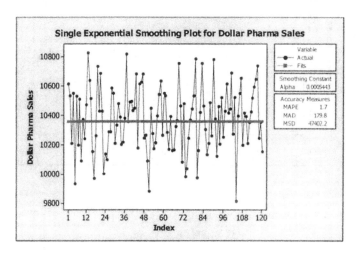

**b.**

| Actual | Forecast $\lambda = 0.1$ | Error $\lambda = 0.1$ | Forecast $\lambda$ = optimal | Error $\lambda$ = optimal |
|---|---|---|---|---|
| 10,210.1 | 10,393.2 | −183.1 | 10,357.2 | −147.1 |
| 10,352.5 | 10,374.9 | −22.4 | 10,357.1 | −4.6 |
| 10,423.8 | 10,372.7 | 51.1 | 10,357.1 | 66.7 |
| 10,519.3 | 10,377.8 | 141.5 | 10,357.1 | 162.2 |
| 10,596.7 | 10,391.9 | 204.8 | 10,357.2 | 239.5 |
| 10,650 | 10,412.4 | 237.6 | 10,357.3 | 292.7 |
| 10,741.6 | 10,436.2 | 305.4 | 10,357.5 | 384.1 |
| 10,246 | 10,466.7 | −220.7 | 10,357.7 | −111.7 |
| 10,354.4 | 10,444.6 | −90.2 | 10,357.6 | −3.2 |
| 10,155.4 | 10,435.6 | −280.2 | 10,357.6 | −202.2 |
| | Stdev→ | 205.9168 | Stdev→ | 197.3727 |

**c.** The data appear to be highly uncorrelated, indicating that the value of $\lambda$ will be very small.

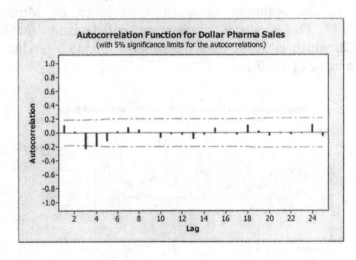

**4.17** Below is a plot of the first differences of the blue and gorgonzola cheese data from Table B.4. The differencing has removed the linear upward trend from the data.

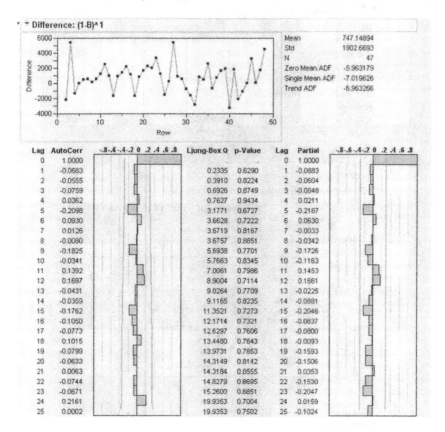

The figure below shows the results of the exponential smoothing on the first differences. The simple exponential smoothing equation can be used directly to forecast the differences. The differenced data can then be converted back to the original data. Another equivalent forecasting method would be to use the IMA(2,1) equation, which is equivalent to simple exponential smoothing on the first differences. This IMA(2,1) equation can be used to forecast the original data.

## JMP Output

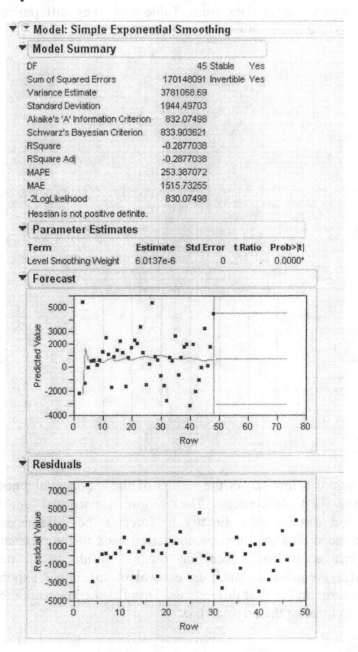

**4.19**   **a.**

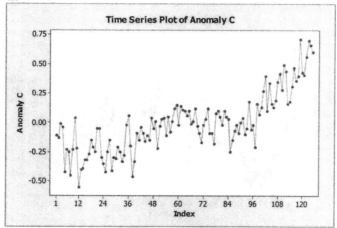

**b.** Using a smoothing constant of $\lambda = 0.2$, the simple exponential smoothing seems to work well for the air temperature anomaly data. This is because the fits seem to not only smooth the data but also capture the trending seen in the data.

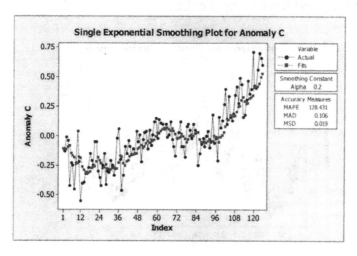

**4.23    a.**

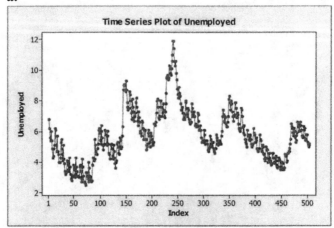

**b.** As seen from the graph below of the simple exponential smoothing for the unemployment data, this procedure appears to work fairly well. It does appear that there may be cyclic trending in the data, in which case a seasonal model might also work well with smoothing and forecasting the data.

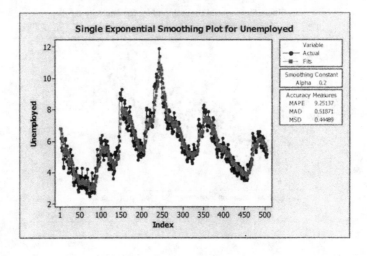

**4.27  a.** The times series plot of the airline miles data verifies that the data are seasonal. See plot.

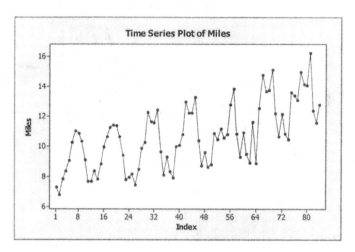

**b.** Winters' multiplicative method appears to do well at capturing the trends of the data; the fits appear to be close to the actual values.

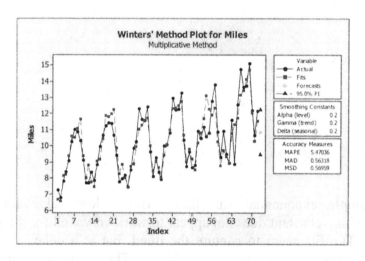

**c.**

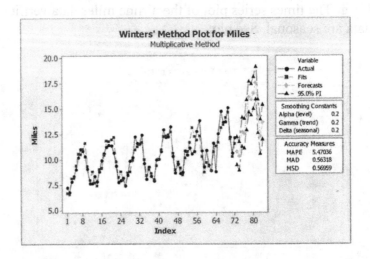

**4.33    a.**

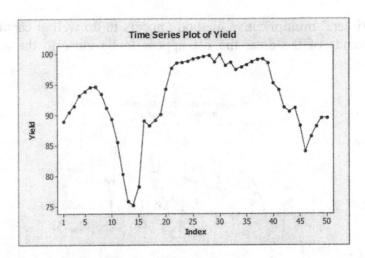

**b.** Single exponential smoothing with a low value for the smoothing constant does not appear to work very well for the yield data. The fits seem to capture the trend, but at a delay, and the mean squared deviation is high. The structure of the autocorrelation function (see exercise 4.35) and the partial autocorrelation function for these data suggests that the data would be best fit by some kind of autoregressive model or autoregressive and moving average model, which will be discussed in the next chapter. If single exponential smoothing were to be used for these data, the smoothing constant should be set to a much higher level to keep up with the volatile trends in the data.

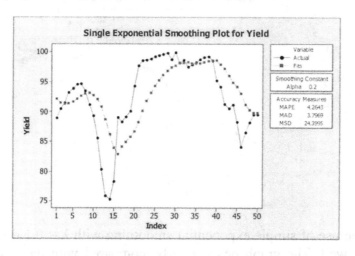

**4.35** The autocorrelation function for the yield data indicates a trend. This would suggest that an autoregressive model might provide a better fit to the data. However, an exponential smoothing model might work if the smoothing constant was chosen to be very high (near or above a value of 1).

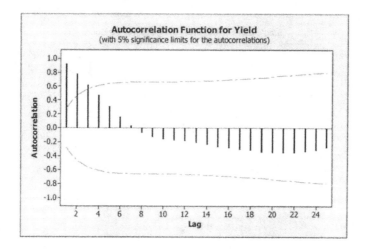

**4.37** **a.** The optimum smoothing constant is $\lambda = 1.92880$.

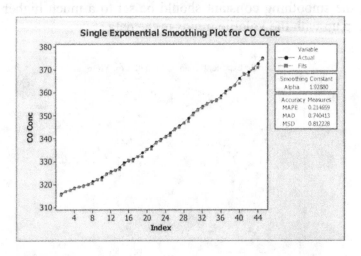

**b.** The use of simple exponential smoothing with $\lambda = 0.1$ does not work well. The graph below can be compared with the graph in part (a). The graph below demonstrates that the fits are very far from the actual data, where the graph in part (a) has the fits directly in line with the actual values. The improvement is best noted by comparing the mean squared deviations (MSDs). The use of the optimum value for $\lambda$ versus a value of 0.1 reduces the MSD from 116.982 to 0.81!

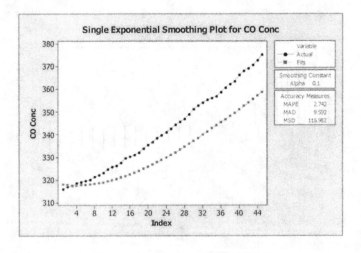

**4.39** An exponential smoothing forecasting procedure for the gross domestic product (GDP) can be created by using Minitab (or other software package) to find the optimum value for $\lambda$, the exponential smoothing constant. In this case the optimum value for $\lambda$ is 1.84. The equation

$$y_{t+1} = \hat{y}_t + 1.84e$$

can then be used to predict new values of GDP.

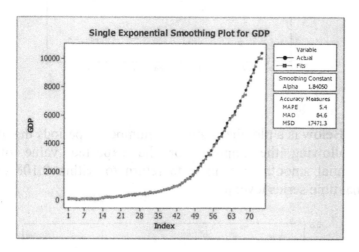

**4.45** Below is a plot of the number of time periods it takes for the expected value of the exponential smoothing statistic to be within $0.10\delta$ of the new time series level $\mu + \delta$. As $\lambda$ approaches 1, the time it takes to be within $0.10\delta$ of the new time series approaches 0. This is intuitive as the exponential smoothed statistic contains an additive portion with a multiple of $1 - \lambda$, which goes to 0 as $\lambda$ goes to 1.

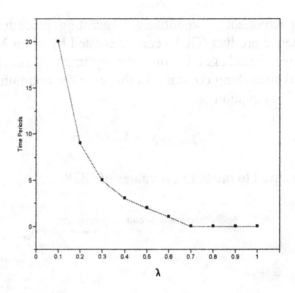

**4.47** Below is a plot that shows the number of periods that it will take following the impulse for the expected value of the exponential smoothing statistic to return to within $0.10\delta$ of the original time series level $\mu$.

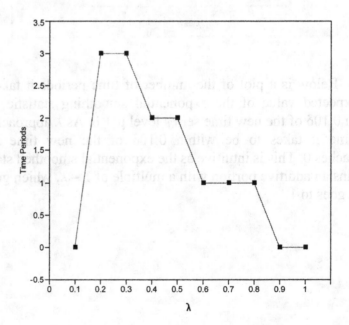

# Chapter 5

# Autoregressive Integrated Moving Average (ARIMA) Models

## Exercises

**5.3** **a.** Below is a times series plot of the data in Table E5.2.

**b.** The sample autocorrelation and partial autocorrelation functions are also shown in the figure below. The sample autocorrelation function has a slightly decaying series and the partial autocorrelation function is significant at lag 1.

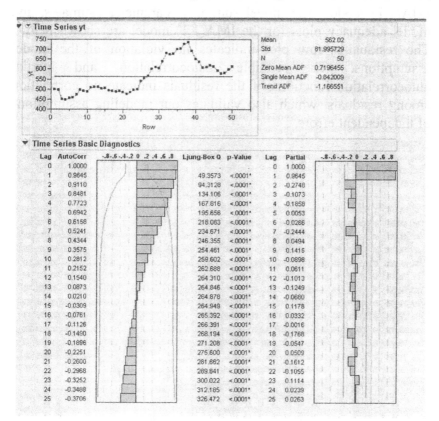

**c.** One appropriate model for these data is the AR(1) model. Another good choice might also be an IMA(1,1), which is an EWMA model. This is because the variogram (not shown here) increases linearly. A comparison of three models from the JMP output is shown in the table below.

| Model | DF | Variance | AIC | SBC | $R^2$ | AIC Rank |
|---|---|---|---|---|---|---|
| AR(1) | 48 | 363.25 | 441.35 | 445.18 | 0.938 | 3 |
| ARIMA(1, 1, 1) | 46 | 295.21 | 421.07 | 426.74 | 0.958 | 2 |
| IMA(1, 1) | 47 | 293.20 | 419.78 | 423.56 | 0.958 | 1 |

The table indicates that the IMA(1,1) model would be a good choice because it has the lowest variance. While there are several model choices that would be adequate, we will show the fits for the IMA(1,1) model. The model summary, parameter estimates, and model adequacy plots for the IMA(1,1) model are shown below. The residual-by-row plot indicates no violation of the model assumptions. The sample autocorrelation and partial autocorrelation functions of the residuals indicate no correlation among residuals, which also validates our modeling assumptions of independent errors.

# JMP Output

▼ Model: IMA(1, 1)
  ▼ Model Summary

| | |
|---|---|
| Variance Estimate | 293.191009 |
| Standard Deviation | 17.1230222 |
| Akaike's 'A' Information Criterion | 419.780209 |
| Schwarz's Bayesian Criterion | 423.563849 |
| RSquare | 0.95797658 |
| RSquare Adj | 0.95708247 |
| MAPE | 2.37377641 |
| MAE | 13.3534523 |
| -2LogLikelihood | 415.780209 |

  ▼ Parameter Estimates

| Term | Lag | Estimate | Std Error | t Ratio | Prob>\|t\| | Constant Estimate |
|---|---|---|---|---|---|---|
| MA1 | 1 | -0.578885 | 0.135294 | -4.26 | <.0001* | 2.59255346 |
| Intercept | 0 | 2.592553 | 3.750443 | 0.69 | 0.4928 | |

  ▶ Forecast
  ▼ Residuals

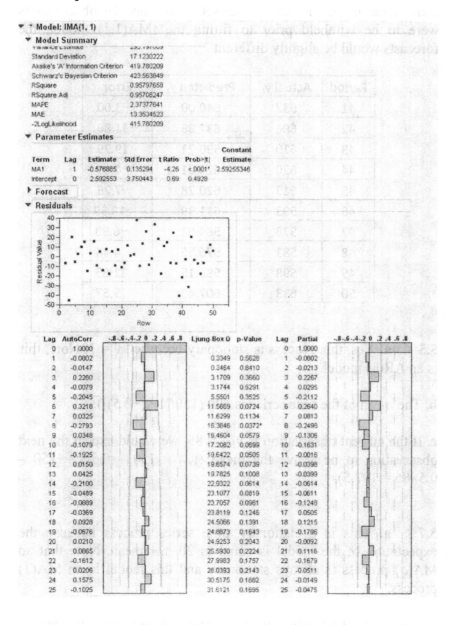

| Lag | AutoCorr | -.8-.6-.4-.2 0 .2 .4 .6 .8 | Ljung-Box Q | p-Value | Lag | Partial | -.8-.6-.4-.2 0 .2 .4 .6 .8 |
|---|---|---|---|---|---|---|---|
| 0 | 1.0000 | | | | 0 | 1.0000 | |
| 1 | -0.0802 | | 0.3349 | 0.5628 | 1 | -0.0802 | |
| 2 | -0.0147 | | 0.3464 | 0.8410 | 2 | -0.0213 | |
| 3 | 0.2260 | | 3.1709 | 0.3660 | 3 | 0.2267 | |
| 4 | -0.0079 | | 3.1744 | 0.5291 | 4 | 0.0295 | |
| 5 | -0.2045 | | 5.5501 | 0.3525 | 5 | -0.2112 | |
| 6 | 0.3218 | | 11.5669 | 0.0724 | 6 | 0.2640 | |
| 7 | 0.0325 | | 11.6299 | 0.1134 | 7 | 0.0813 | |
| 8 | -0.2793 | | 16.3846 | 0.0372* | 8 | -0.2498 | |
| 9 | 0.0348 | | 16.4604 | 0.0579 | 9 | -0.1306 | |
| 10 | -0.1079 | | 17.2082 | 0.0699 | 10 | -0.1631 | |
| 11 | -0.1925 | | 19.6422 | 0.0505 | 11 | -0.0016 | |
| 12 | 0.0150 | | 19.6574 | 0.0739 | 12 | -0.0398 | |
| 13 | 0.0425 | | 19.7825 | 0.1008 | 13 | -0.0399 | |
| 14 | -0.2100 | | 22.9322 | 0.0614 | 14 | -0.0614 | |
| 15 | -0.0489 | | 23.1077 | 0.0819 | 15 | -0.0611 | |
| 16 | -0.0689 | | 23.7057 | 0.0961 | 16 | -0.1248 | |
| 17 | -0.0369 | | 23.8119 | 0.1246 | 17 | 0.0505 | |
| 18 | 0.0928 | | 24.5066 | 0.1391 | 18 | 0.1215 | |
| 19 | -0.0676 | | 24.8873 | 0.1643 | 19 | -0.1796 | |
| 20 | 0.0210 | | 24.9253 | 0.2043 | 20 | -0.0092 | |
| 21 | 0.0665 | | 25.5930 | 0.2224 | 21 | 0.1116 | |
| 22 | -0.1612 | | 27.9983 | 0.1757 | 22 | -0.1679 | |
| 23 | 0.0206 | | 28.0393 | 0.2143 | 23 | -0.0511 | |
| 24 | 0.1575 | | 30.5175 | 0.1682 | 24 | -0.0149 | |
| 25 | -0.1025 | | 31.6121 | 0.1695 | 25 | -0.0475 | |

**d.** The table below contains the one-step-ahead forecasts of the last 10 observations and the forecast errors. Note that these forecasts were made using the entire data set. If the last 10 observations were to be withheld prior to fitting the IMA(1,1) model, the forecasts would be slightly different.

| Period | Actual $y_t$ | Predicted $y_t$ | Error |
|--------|--------------|-----------------|--------|
| 41 | 637 | 640.00 | −3.00 |
| 42 | 606 | 637.86 | −31.86 |
| 43 | 610 | 590.21 | 19.79 |
| 44 | 620 | 624.01 | −4.01 |
| 45 | 613 | 620.28 | −7.28 |
| 46 | 593 | 611.39 | −18.39 |
| 47 | 578 | 584.98 | −6.98 |
| 48 | 581 | 576.56 | 4.44 |
| 49 | 598 | 586.15 | 11.85 |
| 50 | 613 | 607.43 | 5.57 |

**5.5**　　**a.** Yes, this process is stationary because $|\varphi| < 1$. Note, this is an AR(1) model.

**b.** The mean of the time series is 100 (150/[1− (−0.5)]).

**c.** If the current observation is $y_{100} = 85$, we would expect the next observation to be above the mean ($y_{t+1} = 150 - 0.5y_t = 150 - 0.5 \times 85 = 107.5$).

**5.7**　　**a.** This is a stationary time series process because the expectation of the model is equal to 20, the mean. Note that an MA($q$) process is always stationary and this model is an MA(1) process.

**b.** Yes. Note, the MA($q$) process is said to be invertible if it has an absolutely summable infinite AR representation. Also, if the sum of the MA parameters is less than 1, then the process is invertible. Here we have a single MA parameter with value 0.2, whose absolute value is less than 1.

c. The mean of this time series is 20.

d. If the current observation is $y_{100} = 23$, we would expect the next observation to be either above or below the mean. The next observations are dependent on the previous error, and until we know that value, we do not know if the observation will be above or below the mean.


**5.9**    a. An MA(1) model can be used to fit the first differences. The terms in this model are significant and the model diagnostics confirm the modeling assumptions. Note that using an MA(1) model to fit the first differences is equivalent to fitting an IMA(0,1,1) model to the original data and the IMA(0,1,1) model is an EWMA model. The model summary is shown in the figure below.

b. The EWMA model can be used for forecasting off of the original data.

## JMP Output

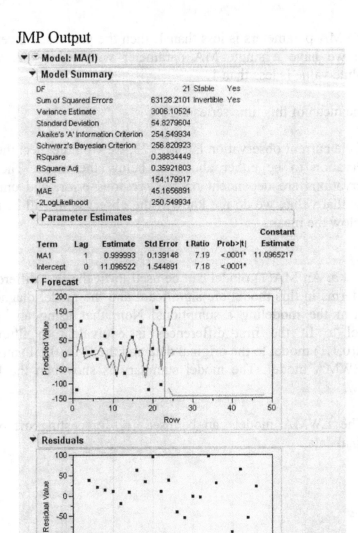

**Model: MA(1)**

**Model Summary**

| | |
|---|---|
| DF | 21 Stable Yes |
| Sum of Squared Errors | 63128.2101 Invertible Yes |
| Variance Estimate | 3006.10524 |
| Standard Deviation | 54.8279604 |
| Akaike's 'A' Information Criterion | 254.549934 |
| Schwarz's Bayesian Criterion | 256.820923 |
| RSquare | 0.38834449 |
| RSquare Adj | 0.35921803 |
| MAPE | 154.179917 |
| MAE | 45.1656891 |
| -2LogLikelihood | 250.549934 |

**Parameter Estimates**

| Term | Lag | Estimate | Std Error | t Ratio | Prob>|t| | Constant Estimate |
|---|---|---|---|---|---|---|
| MA1 | 1 | 0.999993 | 0.139148 | 7.19 | <.0001* | 11.0965217 |
| Intercept | 0 | 11.096522 | 1.544891 | 7.18 | <.0001* | |

**Forecast**

**Residuals**

**5.13**  **a.** There are several models that can be used to forecast the U.S. production of blue and gorgonzola cheeses. One good choice is the IMA(1,1) model. This model provides low variance and high $R^2$ and does not have any violations of the modeling assumptions.

The figure below contains the model summary, parameter estimates, forecast graph, and residual-versus-row values.

### JMP Output

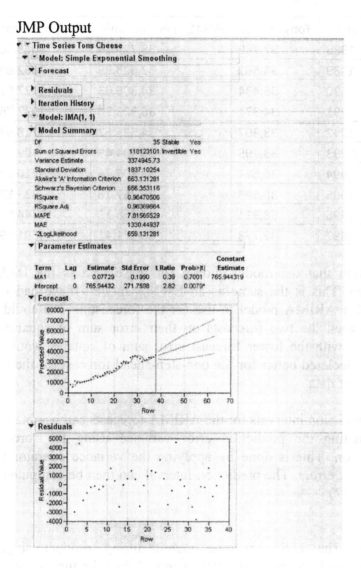

Time Series Tons Cheese
* Model: Simple Exponential Smoothing
  * Forecast
  * Residuals
  * Iteration History
* Model: IMA(1, 1)
  * Model Summary

| DF | 35 | Stable | Yes |
|---|---|---|---|
| Sum of Squared Errors | 118123101 | Invertible | Yes |
| Variance Estimate | 3374945.73 | | |
| Standard Deviation | 1837.10254 | | |
| Akaike's 'A' Information Criterion | 663.131281 | | |
| Schwarz's Bayesian Criterion | 666.353116 | | |
| RSquare | 0.96470506 | | |
| RSquare Adj | 0.96369664 | | |
| MAPE | 7.81565529 | | |
| MAE | 1330.44937 | | |
| -2LogLikelihood | 659.131281 | | |

* Parameter Estimates

| Term | Lag | Estimate | Std Error | t Ratio | Prob>|t| | Constant Estimate |
|---|---|---|---|---|---|---|
| MA1 | 1 | 0.07729 | 0.1990 | 0.39 | 0.7001 | 765.944319 |
| Intercept | 0 | 765.94432 | 271.7598 | 2.82 | 0.0079* | |

* Forecast

* Residuals

**b.** The forecast of the last 10 observations is shown in the table below.

| Year | Tons Cheese | IMA(1,1) Prediction | IMA(1,1) Error |
|------|-------------|---------------------|----------------|
| 1988 | 37,789 | 35,725.14 | 2063.86 |
| 1989 | 34,561 | 37,623.89 | −3062.89 |
| 1990 | 36,434 | 34,806.03 | 1627.97 |
| 1991 | 34,371 | 36,303.76 | −1932.76 |
| 1992 | 33,307 | 34,525.62 | −1218.62 |
| 1993 | 33,295 | 33,404.49 | −109.49 |
| 1994 | 36,514 | 33,303.76 | 3210.24 |
| 1995 | 36,593 | 36,257.18 | 335.82 |
| 1996 | 38,311 | 36,566.13 | 1744.87 |
| 1997 | 42,773 | 38,171.41 | 4601.59 |

**c.** Note that the model chosen for these data is the IMA(1,1) model. This is the same as an EWMA model. If we had picked another ARIMA model to use for the forecasting, we could have compared the two forecasts on their error sum of squares. The model with the lower forecast error sum of squares would have been declared better for the one-step-ahead forecast for the last 10 years of data.

**d.** Prediction intervals for the ARIMA forecasts can be obtained by estimating the prediction error variance using the forecasting equation. This is done by applying the variance operator to the forecast errors. The prediction interval can then be computed as in Eq.(5.82).

**5.15** The U.S. beverage manufacturing product shipments appear to be a seasonal ARIMA model. See the time series graph and ACF and PACF plots in the figure below.

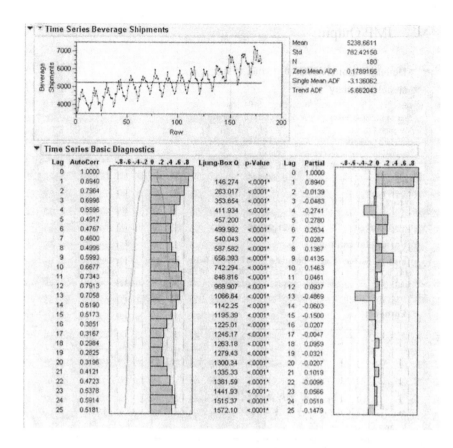

For these time series data, an appropriate model would be the seasonal ARIMA$(0,1,1)\times(0,1,1)_{12}$. The model summary output and parameter estimates are shown below. To find prediction intervals for future forecasts, find the prediction variance by applying the variance operator on the forecast model. This prediction variance can then be used for the prediction interval calculation.

## JMP Output

### ▼ ⫶ Model: Seasonal ARIMA(0, 1, 1)(0, 1, 1)12

#### ▼ Model Summary

| | | | |
|---|---|---|---|
| DF | 164 | Stable | Yes |
| Sum of Squared Errors | 3080620.76 | Invertible | Yes |
| Variance Estimate | 18784.273 | | |
| Standard Deviation | 137.055729 | | |
| Akaike's 'A' Information Criterion | 2128.64164 | | |
| Schwarz's Bayesian Criterion | 2137.99562 | | |
| RSquare | 0.96526554 | | |
| RSquare Adj | 0.96484195 | | |
| MAPE | 1.98523061 | | |
| MAE | 105.544056 | | |
| -2LogLikelihood | 2122.64164 | | |

#### ▼ Parameter Estimates

| Term | Factor | Lag | Estimate | Std Error | t Ratio | Prob>|t| | Constant |
|---|---|---|---|---|---|---|---|
| | | | | | | | Estimate |
| MA1,1 | 1 | 1 | 0.63728150 | 0.060496 | 10.53 | <.0001* | |
| MA2,12 | 2 | 12 | 0.69155842 | 0.078635 | 8.79 | <.0001* | 0.99896836 |
| Intercept | 1 | 0 | 0.99896836 | 1.446288 | 0.69 | 0.4907 | |

#### ▼ Forecast

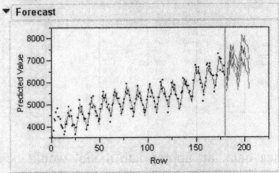

#### ▼ Residuals

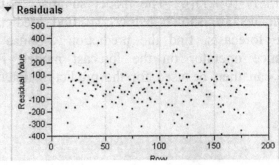

**5.17**    Below is a graph with the EWMA model fits to the data in Table B.6. The optimum smoothing constant λ = 0.33.

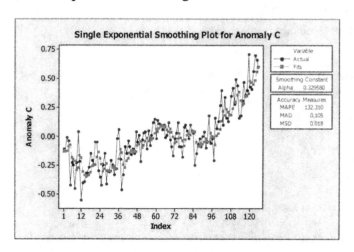

The ARIMA model chosen as appropriate for these data is an IMA(1,2) model. The model fits are displayed below. The IMA(1,2) model output indicates that the variances of the fits for this model are smaller than the fits to the single exponential smoothing model [IMA(1,1)]. This indicates that the IMA(1,2) is a more desirable model to use for forecasting.

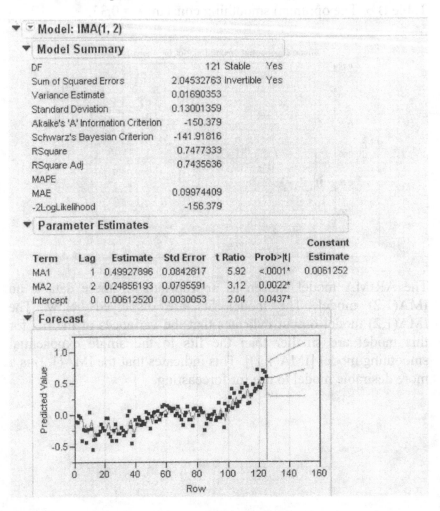

▼ ▽ Model: IMA(1, 2)
  ▼ Model Summary

| | |
|---|---|
| DF | 121 Stable Yes |
| Sum of Squared Errors | 2.04532763 Invertible Yes |
| Variance Estimate | 0.01690353 |
| Standard Deviation | 0.13001359 |
| Akaike's 'A' Information Criterion | -150.379 |
| Schwarz's Bayesian Criterion | -141.91816 |
| RSquare | 0.7477333 |
| RSquare Adj | 0.7435636 |
| MAPE | |
| MAE | 0.09974409 |
| -2LogLikelihood | -156.379 |

  ▼ Parameter Estimates

| Term | Lag | Estimate | Std Error | t Ratio | Prob>\|t\| | Constant Estimate |
|---|---|---|---|---|---|---|
| MA1 | 1 | 0.49927896 | 0.0842817 | 5.92 | <.0001* | 0.0061252 |
| MA2 | 2 | 0.24856193 | 0.0795591 | 3.12 | 0.0022* | |
| Intercept | 0 | 0.00612520 | 0.0030053 | 2.04 | 0.0437* | |

  ▼ Forecast

**5.19** Below is a graph with the EWMA model fits to the data in Table B.7. The optimum smoothing constant $\lambda = 1.08$. The fits appear to follow closely the actual data.

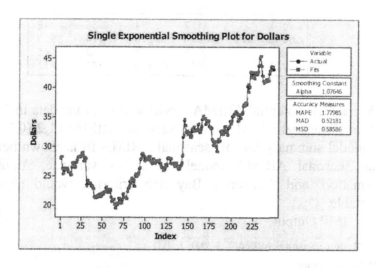

The autocorrelation and partial autocorrelation functions indicate that there is an autoregressive component at work. Two models were chosen as appropriate for these data. One was the EWMA model and the other an ARIMA(1,1,1) model. The model summary for the ARIMA(1,1,1) fit is displayed below. The mean squared error for this model is 0.588, where the mean squared error for the single exponential smoothing model was 0.586. These numbers are very similar and both models would be appropriate.

Minitab Output

```
Final Estimates of Parameters

Type           Coef  SE Coef       T       P
AR    1     -0.6908   0.2675   -2.58   0.010
MA    1     -0.7669   0.2363   -3.25   0.001
Constant   0.10328  0.08625    1.20   0.232


Differencing: 1 regular difference
Number of observations:  Original series 248,
after differencing 247
Residuals:    SS =  143.589 (backforecasts
excluded)
              MS =  0.588  DF = 244


Modified Box-Pierce (Ljung-Box) Chi-Square
statistic
```

| Lag | 12 | 24 | 36 | 48 |
|-----------|-------|-------|-------|-------|
| Chi-Square | 4.6 | 12.0 | 24.4 | 39.3 |
| DF | 9 | 21 | 33 | 45 |
| P-Value | 0.871 | 0.939 | 0.862 | 0.711 |

**5.25** **a.** One potential ARIMA model to use for the data in Table B.10 is a seasonal ARIMA model such as $ARIMA(0,1,2)(0,1,1)_{12}$. The model summary for this seasonal ARIMA fit is shown below. Other seasonal ARIMA models with low variance, Akaike's information, and Schwarz's Bayesian criterion would also be acceptable.

JMP Output

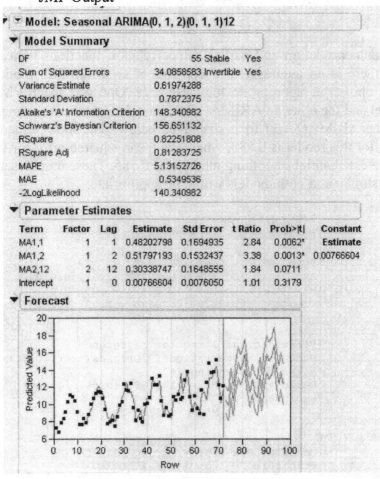

**Model: Seasonal ARIMA(0, 1, 2)(0, 1, 1)12**

**Model Summary**

| | | | |
|---|---|---|---|
| DF | 55 | Stable | Yes |
| Sum of Squared Errors | 34.0858583 | Invertible | Yes |
| Variance Estimate | 0.61974288 | | |
| Standard Deviation | 0.7872375 | | |
| Akaike's 'A' Information Criterion | 148.340982 | | |
| Schwarz's Bayesian Criterion | 156.651132 | | |
| RSquare | 0.82251808 | | |
| RSquare Adj | 0.81283725 | | |
| MAPE | 5.13152726 | | |
| MAE | 0.5349536 | | |
| -2LogLikelihood | 140.340982 | | |

**Parameter Estimates**

| Term | Factor | Lag | Estimate | Std Error | t Ratio | Prob>\|t\| | Constant |
|------|--------|-----|----------|-----------|---------|----------|----------|
| MA1,1 | 1 | 1 | 0.48202798 | 0.1694935 | 2.84 | 0.0062* | **Estimate** |
| MA1,2 | 1 | 2 | 0.51797193 | 0.1532437 | 3.38 | 0.0013* | 0.00766604 |
| MA2,12 | 2 | 12 | 0.30338747 | 0.1648555 | 1.84 | 0.0711 | |
| Intercept | 1 | 0 | 0.00766604 | 0.0076050 | 1.01 | 0.3179 | |

**Forecast**

**b.** A lower sum of squares of the errors is a good indication that the predictions match the actual data, which is the goal of the forecast.

**5.31** An appropriate ARIMA model for the ice cream and frozen yogurt sales data in Table B.13 would be an IMA(1,1) model. Note that the IMA model is equivalent to the EWMA (single exponential smoothing method). The plot below shows the actual and fitted data along with the optimal smoothing constant $\lambda = 1.37$.

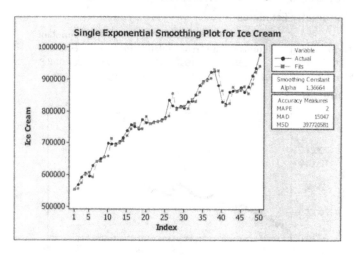

To make a comparison, the model summary from JMP is also displayed below. Note that after manipulation the optimum value for $\lambda$ from the JMP output is 1.27, which is a tenth less than the Minitab output shown above. This is due to the way in which the software packages estimate the starting estimate. Prediction intervals can be computed by obtaining estimates of prediction variance from the forecasting equation and substituting this variance estimate into the standard prediction interval equation.

## JMP Output

**▼ ▼ Model: IMA(1, 1)**

**▼ Model Summary**

| | |
|---|---|
| DF | 47 Stable Yes |
| Sum of Squared Errors | 1.7656e+10 Invertible Yes |
| Variance Estimate | 375658563 |
| Standard Deviation | 19381.9133 |
| Akaike's 'A' Information Criterion | 1108.55454 |
| Schwarz's Bayesian Criterion | 1112.33818 |
| RSquare | 0.96429114 |
| RSquare Adj | 0.96353138 |
| MAPE | 1.73533639 |
| MAE | 13808.2728 |
| -2LogLikelihood | 1104.55454 |

**▼ Parameter Estimates**

| Term | Lag | Estimate | Std Error | t Ratio | Prob>\|t\| | Constant Estimate |
|---|---|---|---|---|---|---|
| MA1 | 1 | -0.269 | 0.159 | -1.69 | 0.0975 | 8862.01406 |
| Intercept | 0 | 8862.014 | 3428.149 | 2.59 | 0.0129* | |

**▼ Forecast**

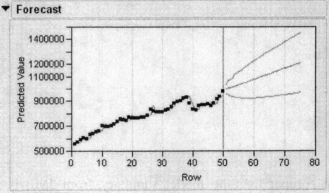

**5.35** The time series plot, autocorrelation, partial autocorrelation, and variogram plots indicate that an IMA(1,1) model would be a good choice. See the plots below.

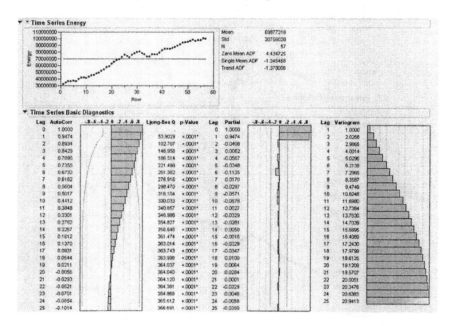

In fact, IMA(1,1) is an appropriate model. The model summary and residuals are shown below. Prediction intervals can be computed by calculating the prediction variance of the IMA(1,1) equation. To do this, apply the variance operator to the forecasting equation. Use the estimates of prediction variance at each time $t$ in the future in the prediction interval equation to obtain the range of the predictions.

## JMP Output

▼ Model Summary

| | |
|---|---|
| Sum of Squared Errors | 1.5288e+14 Invertible Yes |
| Variance Estimate | 2.8311e+12 |
| Standard Deviation | 1682600.18 |
| Akaike's 'A' Information Criterion | 1766.57533 |
| Schwarz's Bayesian Criterion | 1770.62603 |
| RSquare | 0.99337404 |
| RSquare Adj | 0.99325133 |
| MAPE | 2.01565029 |
| MAE | 1316170.16 |
| -2LogLikelihood | 1762.57533 |

▼ Parameter Estimates

| Term | Lag | Estimate | Std Error | t Ratio | Prob>|t| | Constant Estimate |
|---|---|---|---|---|---|---|
| MA1 | 1 | -0.2696778 | 0.1295013 | -2.08 | 0.0421* | 1207585.46 |
| Intercept | 0 | 1207585 | 279323.4 | 4.32 | <.0001* | |

▼ Forecast

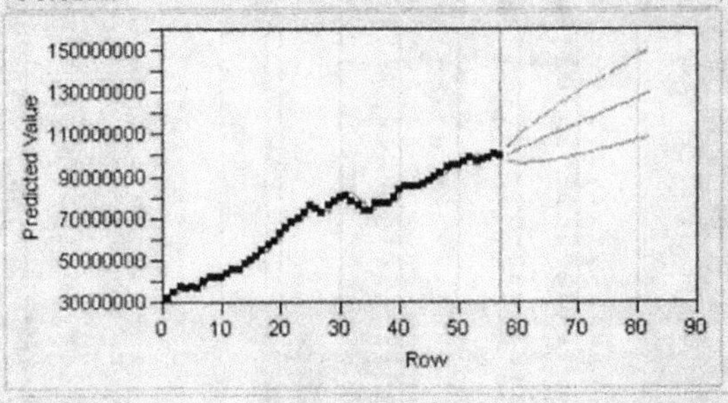

▼ Residuals

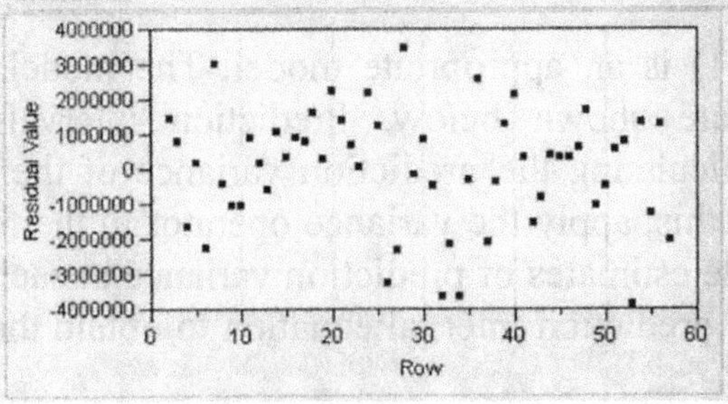

**5.39**  **a.** A plot of the forecasts starting from the origin 100 is displayed in the graph below.

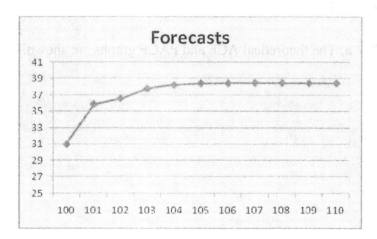

**b.** The shape of the forecast function from the model increases slightly for the first few time periods and then becomes level after period 104.

**c.** A plot with the updated observation from time period 101 is shown.

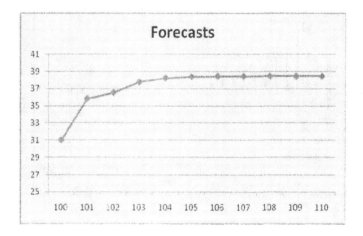

**d.** The 95% prediction interval on the forecast of period 101 made at the end of period 100 is [33.08,38.62]. This is calculated from $\hat{y}_t \pm Z_{\alpha/2}\hat{\sigma}_e$.

**5.47    a.** The theoretical ACF and PACF graphs are shown.

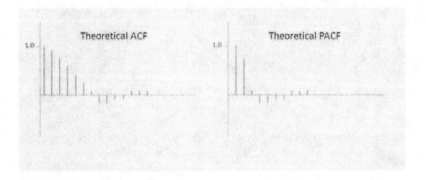

**b.** The sample ACF and PACF from 50 realizations of the AR(1) process are shown below. These are almost identical to the theoretical graphs in part (a).

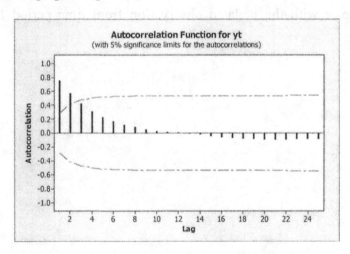

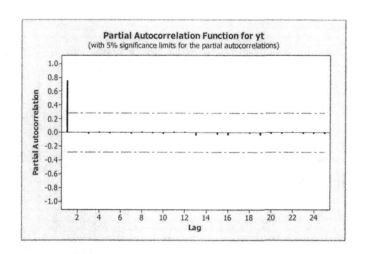

**c.** Increasing the sample size did not impact the agreement between the sample and theoretical ACF and PACF values because there was already complete agreement at a sample size of 50. If the variance of the error term was larger, more samples would be required to achieve this agreement.

**5.51** **a.** The theoretical ACF should slowly decay but alternate from negative to positive. The theoretical PACF should have a single negative spike.

**b.** Displayed below are the ACF and PACF from 50 realizations of the ARMA(1,1) model.

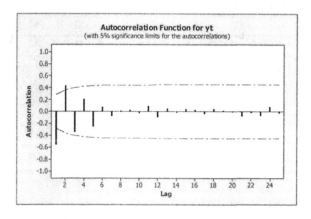

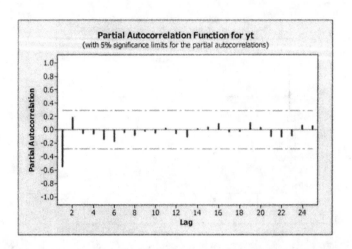

**c.** Displayed below are the ACF and PACF from 200 realizations of the ARMA(1,1) model. The reliability of the results of a time series model can often increase as the sample size is increased.

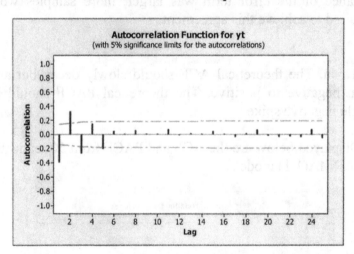

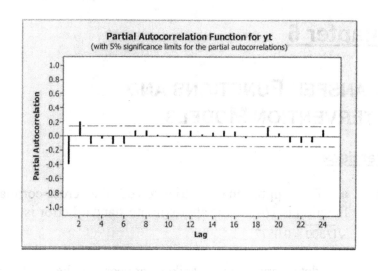

# Chapter 6

# TRANSFER FUNCTIONS AND INTERVENTION MODELS

## Exercises

**6.1**    **a.** The approximate variance of the cross-correlation function is $1/N$, so the approximate standard error is
$$\sqrt{1/300} = 0.0577$$

To determine which spikes appear to be significant, construct approximate 95% confidence intervals (CIs) for each lag and see which ones do not contain zero.

| Lag, $j$ | 0 | 1 | 2 | 3 | 4 | 5 | 6 | 7 | 8 | 9 | 10 |
|---|---|---|---|---|---|---|---|---|---|---|---|
| $r_{\alpha\beta}(j)$ | 0.01 | 0.03 | -0.03 | -0.25 | -0.35 | -0.51 | -0.30 | -0.15 | -0.02 | 0.09 | -0.01 |
| $v_{\alpha\beta}(j)$ | 0.02 | 0.06 | -0.06 | -0.50 | -0.70 | -1.02 | -0.60 | -0.30 | -0.04 | 0.18 | -0.02 |
| Lower CI | -0.11 | -0.09 | -0.15 | -0.37 | -0.47 | -0.63 | -0.42 | -0.27 | -0.14 | -0.03 | -0.13 |
| Upper CI | 0.13 | 0.15 | 0.09 | -0.13 | -0.23 | -0.39 | -0.18 | -0.03 | 0.10 | 0.21 | 0.11 |

The CIs for lags 3, 4, 5, 6, and 7 do not contain zero, so these spikes appear to be significant.

**b.** The estimate for the impulse response function is below.

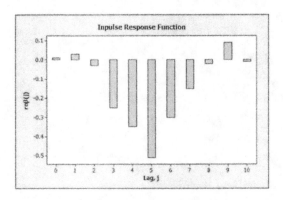

There is a lag of 3, and from Table 6.1, this appears to follow closest to a (2, 2) model. Therefore, the tentative form of the transfer function model is (3, 2, 2).

**6.2** The form of the model is

$$y_t = \frac{w_0 - w_1 B - w_2 B^2}{1 - \delta_1 B - \delta_2 B^2} x_{t-3}$$

From Eq. (6.12), we have

$$v_0 = v_1 = v_2 = 0$$
$$v_3 = w_0$$
$$v_4 = \delta_1 v_3 + \delta_2 v_2 - w_1$$
$$v_5 = \delta_1 v_4 + \delta_2 v_3 - w_2$$
$$v_6 = \delta_1 v_5 + \delta_2 v_4$$
$$v_7 = \delta_1 v_6 + \delta_2 v_5$$

So,
$$\hat{w}_0 = -0.5$$
$$-0.7 = -0.5\delta_1 - w_1$$
$$-1.02 = -0.7\delta_1 - 0.5\delta_2 - w_2$$
$$-0.6 = -1.02\delta_1 - 0.7\delta_2$$
$$-0.3 = -0.6\delta_1 - 1.02\delta_2$$

Solve the last four equations simultaneously:

$$\hat{w}_0 = -0.5, \ \hat{w}_1 = 0.376, \ \hat{w}_2 = 0.610, \ \hat{\delta}_1 = 0.648, \ \hat{\delta}_2 = -0.087$$

**6.5** The form of the model is

$$y_t = \frac{w_0}{1 - \delta_1 B} x_{t-2}$$

From Eq. (6.12), we have

$$v_0 = v_1 = 0, \ v_2 = w_0, \ v_3 = \delta_1 w_0, \ v_j = \delta_1 v_{j-1}, \ j > 3$$

Note that $|\delta_1| < 1$ for the transfer function to achieve stability. Thus, the impulse response function will go to zero asymptotically.

**6.7** The form of the model is

$$y_t = \frac{w_0 - w_1 B}{1 - \delta_1 B} x_{t-2}$$

From Eq. (6.12), we have

$$v_0 = v_1 = 0, \ v_2 = w_0, \ v_3 = \delta_1 w_0 - w_1, \ v_j = \delta_1 v_{j-1}, \ j > 3$$

Note that $|\delta_1| < 1$ for the transfer function to achieve stability. Thus, the impulse response function will go to zero asymptotically.

**6.13** A sketch of the impulse response function is below.

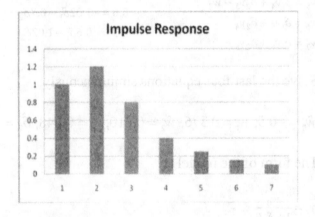

**6.15** **a.** $b = 0$ (no lag), $r = 1$ (one term in the denominator), and $s = 2$ (two terms in the numerator).

**b.** There are no terms in the numerator and two in the denominator, so the errors follow an AR(2) model.

**c.** Since there is no lag, the carbon dioxide concentration in the output is impacted immediately.

**6.17**   The JMP output is below.

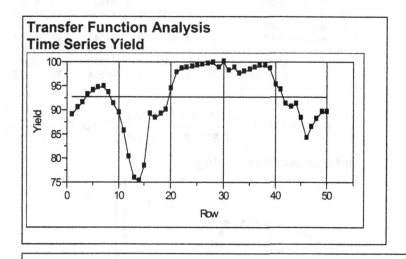

## Transfer Function Analysis
## Time Series Yield

| | |
|---|---|
| Mean | 92.668 |
| Std | 6.3175768 |
| N | 50 |
| Zero Mean ADF | -0.065322 |
| Single Mean ADF | -1.381097 |
| Trend ADF | -1.247108 |

## Model Comparison

| Model | DF | Var | AIC | SBC | R^2 | -2LogLH | MAPE | MAE |
|---|---|---|---|---|---|---|---|---|
| Model | 41 | 4.375 | 215.90 | 229.00 | 0.910 | 201.899 | 1.416 | 1.2743 |

## Transfer Function Model (1)
## Model Summary

| | |
|---|---|
| DF | 41 |
| Sum of Squared Errors | 179.390284 |
| Variance Estimate | 4.37537331 |
| Standard Deviation | 2.0917393 |
| Akaike's 'A' Information Criterion | 215.899215 |
| Schwarz's Bayesian Criterion | 228.997622 |
| RSquare | 0.91010658 |
| RSquare Adj | 0.89756331 |
| MAPE | 1.41618272 |
| MAE | 1.27428723 |
| -2LogLikelihood | 201.899215 |

Failed: Cannot Decrease Objective Function

## Parameter Estimates

| Variable | Term | Factor | Lag | Estimate | Std Error | t Ratio | Prob>|t| |
|---|---|---|---|---|---|---|---|
| Temperature | Num0,0 | 0 | 0 | -0.51083 | 0.151903 | -3.36 | 0.0017 |
| Temperature | Num1,1 | 1 | 1 | -0.52234 | 0.149853 | -3.49 | 0.0012 |
| Temperature | Den1,1 | 1 | 1 | 0.87619 | 0.125905 | 6.96 | <.0001 |
| Temperature | Den1,2 | 1 | 2 | -0.69760 | 0.201143 | -3.47 | 0.0012 |
| Yield | AR1,1 | 1 | 1 | 0.89070 | 0.059204 | 15.04 | <.0001 |
| Yield | MA1,1 | 1 | 1 | -0.41711 | 0.116202 | -3.59 | 0.0009 |
|  | Intercept | 0 | 0 | 90.17163 | 3.799294 | 23.73 | <.0001 |

$$\text{Yield}_t = 90.1716 + \frac{\left(-0.5108 + 0.5223B\right)}{\left(1 - 0.8762B + 0.6976B^2\right)}\text{Temp}_t + \frac{\left(1 + 0.4171B\right)}{\left(1 - 0.8907B\right)}\varepsilon_t$$

## Interactive Forecasting

## Input Time Series Panel
## Input Series: Temperature

Notice that all of the parameters are statistically significant. The AIC, SBC, and $R^2$ values indicate a good model.

**6.21** **a.** The time series plot is given below.

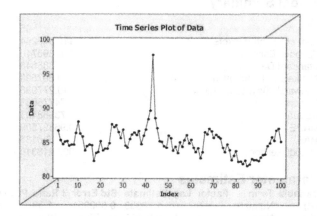

**b.** The outlier appears to be observation #43 (97.85). The JMP output is below.

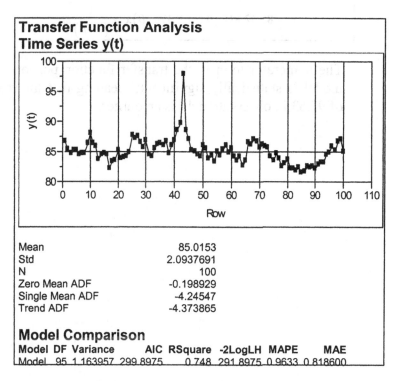

**Transfer Function Analysis**
**Time Series y(t)**

| Mean | 85.0153 |
|---|---|
| Std | 2.0937691 |
| N | 100 |
| Zero Mean ADF | -0.198929 |
| Single Mean ADF | -4.24547 |
| Trend ADF | -4.373865 |

**Model Comparison**

| Model | DF | Variance | AIC | RSquare | -2LogLH | MAPE | MAE |
|---|---|---|---|---|---|---|---|
| Model | 95 | 1.163957 | 299.8975 | 0.748 | 291.8975 | 0.9633 | 0.818600 |

## Transfer Function Model (1)
## Model Summary

| | |
|---|---|
| DF | 95 |
| Sum of Squared Errors | 110.575264 |
| Variance Estimate | 1.16395719 |
| Standard Deviation | 1.07886848 |
| Akaike's 'A' Information Criterion | 299.897508 |
| Schwarz's Bayesian Criterion | 310.277988 |
| RSquare | 0.74776633 |
| RSquare Adj | 0.73988403 |
| MAPE | 0.96327509 |
| MAE | 0.8186 |
| -2LogLikelihood | 303.218316 |

## Parameter Estimates

| Variable | Term | Factor | Lag | Estimate | Std Error | t Ratio | Prob>\|t\| |
|---|---|---|---|---|---|---|---|
| x(t) | Num0,0 | 0 | 0 | 1.20333 | 0.862951 | 1.39 | 0.1664 |
| x(t) | Num1,1 | 1 | 1 | -9.37667 | 0.862931 | -10.87 | <.0001 |
| y(t) | AR1,1 | 1 | 1 | 1.00000 | 2.5451e-6 | 392906 | 0.0000 |
| | Intercept | 0 | 0 | 85.32000 | 1.056848 | 80.73 | <.0001 |

$$y_t = 85.32 + \left(1.2033 + 9.3767B\right)x_t + \frac{1}{\left(1-B\right)}\varepsilon_t$$

The numerator term in the transfer function portion of the model is statistically significant, meaning that the reading of 97.63 in observation 43 is an outlier.

# Chapter 7

# SURVEY OF OTHER FORECASTING METHODS

## Exercises

**7.1** The AR(2) model can be represented as

$$y_t = \phi_1 y_{t-1} + \phi_2 y_{t-2} + \varepsilon_t$$

The state equation can be expressed as

$$\mathbf{X}_t = \mathbf{AX}_{t-1} + \mathbf{G}a_t$$
$$= [\phi_1 \ \phi_2][y_{t-1} \ y_{t-2}]^T + 1\varepsilon_t$$

The observation equation can be expressed as
$$y_t = \mathbf{h}'_t \mathbf{X}_t + \varepsilon_t$$
$$= \phi_1 y_{t-1} + \phi_2 y_{t-2} + \varepsilon_t$$

**7.3** $\hat{\mathbf{p}}_k(T) = \lambda \mathbf{u}_k(T) + (1 - \lambda)\hat{\mathbf{p}}_k(T-1)$

$$
\mathbf{u}(T) = \begin{bmatrix} 0 \\ 0 \\ 0 \\ 0 \\ 0 \\ 0 \\ 0 \\ 0 \\ 0 \\ 1 \\ 0 \end{bmatrix}
\qquad
\hat{\mathbf{p}}_k(T) = 0.1 \begin{bmatrix} 0 \\ 0 \\ 0 \\ 0 \\ 0 \\ 0 \\ 0 \\ 0 \\ 0 \\ 1 \\ 0 \end{bmatrix} + 0.9 \begin{bmatrix} 0.02 \\ 0.03 \\ 0.04 \\ 0.05 \\ 0.08 \\ 0.09 \\ 0.12 \\ 0.17 \\ 0.21 \\ 0.11 \\ 0.08 \end{bmatrix} = \begin{bmatrix} 0.018 \\ 0.027 \\ 0.036 \\ 0.045 \\ 0.072 \\ 0.081 \\ 0.108 \\ 0.153 \\ 0.189 \\ 0.199 \\ 0.072 \end{bmatrix}
\qquad
F(\mathbf{B}_k) = \begin{bmatrix} 0.018 \\ 0.045 \\ 0.081 \\ 0.126 \\ 0.198 \\ 0.279 \\ 0.387 \\ 0.540 \\ 0.729 \\ 0.928 \\ 1.000 \end{bmatrix}
$$

$$\hat{F}_{0.70} = \frac{[0.729 - 0.7]40 + [0.7 - 0.540]50}{0.729 - 0.540} = 48.47$$

**7.5**

$$\mathbf{u}(T) = \begin{bmatrix} 0 \\ 0 \\ 0 \\ 0 \\ 0 \\ 0 \\ 0 \\ 0 \\ 0 \\ 1 \\ 0 \end{bmatrix} \quad \hat{\mathbf{p}}_k(T) = 0.4 \begin{bmatrix} 0 \\ 0 \\ 0 \\ 0 \\ 0 \\ 0 \\ 0 \\ 0 \\ 0 \\ 1 \\ 0 \end{bmatrix} + 0.6 \begin{bmatrix} 0.02 \\ 0.03 \\ 0.04 \\ 0.05 \\ 0.08 \\ 0.09 \\ 0.12 \\ 0.17 \\ 0.21 \\ 0.11 \\ 0.08 \end{bmatrix} = \begin{bmatrix} 0.012 \\ 0.018 \\ 0.024 \\ 0.030 \\ 0.048 \\ 0.054 \\ 0.072 \\ 0.102 \\ 0.126 \\ 0.466 \\ 0.048 \end{bmatrix} \quad F(\mathbf{B}_k) = \begin{bmatrix} 0.012 \\ 0.030 \\ 0.054 \\ 0.084 \\ 0.132 \\ 0.186 \\ 0.258 \\ 0.360 \\ 0.486 \\ 0.952 \\ 1.000 \end{bmatrix}$$

$$\hat{F}_{0.70} = \frac{[0.952 - 0.7]50 + [0.7 - 0.486]55}{0.952 - 0.486} = 52.30$$

Increasing $\lambda$ increased the estimate of the 70th percentile.

**7.7**

$$k^* = \frac{\sigma_2^2 - \rho\sigma_1\sigma_2}{\sigma_1^2 + \sigma_2^2 - 2\rho\sigma_1\sigma_2}$$

$$= \frac{25 - (-0.75)(\sqrt{10})(\sqrt{25})}{10 + 25 - 2(-0.75)(\sqrt{10})(\sqrt{25})}$$

$$= \frac{36.8585}{66.6228} = 0.5532$$

$$\hat{Y}_{t+\tau} = 0.5532\hat{y}_{1,t+\tau} + 0.4468\hat{y}_{2,t+\tau}$$

$$\text{Min Var}[e^c_{t+\tau}] = \frac{\sigma_1^2 \sigma_2^2 (1 - \rho^2)}{\sigma_1^2 + \sigma_2^2 - 2\rho\sigma_1\sigma_2}$$

$$= \frac{(10)(25)(1 - 0.5625)}{10 + 25 - 2(-0.75)(\sqrt{10})(\sqrt{25})}$$

$$= 1.64$$

**7.9**

$$k^* = \frac{\sigma_2^2 - \rho\sigma_1\sigma_2}{\sigma_1^2 + \sigma_2^2 - 2\rho\sigma_1\sigma_2}$$

$$= \frac{16 - (-0.3)(\sqrt{8})(\sqrt{16})}{8 + 16 - 2(-0.3)(\sqrt{8})(\sqrt{16})}$$

$$= \frac{19.3941}{30.7882} = 0.6299$$

$$\hat{Y}_{t+\tau} = 0.6299\hat{y}_{1,t+\tau} + 0.3701\hat{y}_{2,t+\tau}$$

$$\text{Min Var}[e^c_{t+\tau}] = \frac{\sigma_1^2 \sigma_2^2 (1 - \rho^2)}{\sigma_1^2 + \sigma_2^2 - 2\rho\sigma_1\sigma_2}$$

$$= \frac{(8)(16)(0.91)}{8 + 16 - 2(-0.2)(\sqrt{8})(\sqrt{16})}$$

$$= 3.7833$$

**7.11** From Exercise 7.8, $k^* = 0.5512$ and $\text{MinVar}[\hat{y}_{t+\tau}] = 5.1580$.

$$k^* = \frac{\sigma_2^2 - \rho\sigma_1\sigma_2}{\sigma_1^2 + \sigma_2^2 - 2\rho\sigma_1\sigma_2}$$

$$= \frac{20 - (0.4)(\sqrt{15})(\sqrt{20})}{15 + 20 - 2(0.4)(\sqrt{15})(\sqrt{20})}$$

$$= 0.6182$$

$$\hat{Y}_{t+\tau} = 0.6182\hat{y}_{1,t+\tau} + 0.3818\hat{y}_{2,t+\tau}$$

$$\text{Min Var}[e_{t+\tau}^c] = \frac{\sigma_1^2 \sigma_2^2 (1 - \rho^2)}{\sigma_1^2 + \sigma_2^2 - 2\rho\sigma_1\sigma_2}$$

$$= \frac{(15)(20)(0.84)}{15 + 20 - 2(0.4)(\sqrt{15})(\sqrt{20})}$$

$$= 11.9185$$

The weight on the first forecast increases, but the variance of the combined forecast also increases.

**7.13**

$$\mathbf{k} = \frac{\sum_{t+\tau}^{-1}(T)\mathbf{1}}{\mathbf{1}'\sum_{t+\tau}^{-1}(T)\mathbf{1}}$$

$$= \frac{\begin{bmatrix} 27/166 & 11/83 & 4/83 \\ & 73/249 & 19/249 \\ \text{sym.} & & 22/249 \end{bmatrix}\begin{bmatrix} 1 \\ 1 \\ 1 \end{bmatrix}}{\begin{bmatrix} 1 \\ 1 \\ 1 \end{bmatrix}'\begin{bmatrix} 27/166 & 11/83 & 4/83 \\ & 73/249 & 19/249 \\ \text{sym.} & & 22/249 \end{bmatrix}\begin{bmatrix} 1 \\ 1 \\ 1 \end{bmatrix}}$$

$$= \frac{\begin{bmatrix} 57/166 & 125/249 & 53/249 \end{bmatrix}}{527/498}$$

$$= \begin{bmatrix} 171/527 & 250/527 & 106/527 \end{bmatrix}'$$

$$\approx \begin{bmatrix} 0.324478 & 0.474383 & 0.201139 \end{bmatrix}'$$

The formula for the variance of the combined forecast can be derived similarly to the procedure described on p. 366. The variance of the combined forecast for this problem is 0.945.

**7.15** If we use estimates of the variance and correlation between the forecast errors, we find

$$\hat{\sigma}_1 = 1.542, \ \hat{\sigma}_2 = 5.516, \ \hat{\rho} = -0.509$$

$$
\begin{aligned}
k^* &= \frac{\sigma_2^2 - \rho\sigma_1\sigma_2}{\sigma_1^2 + \sigma_2^2 - 2\rho\sigma_1\sigma_2} \\
&= \frac{5.516 - (-0.509)\left(\sqrt{1.542}\right)\left(\sqrt{5.516}\right)}{1.542 + 5.516 - 2(-0.509)\left(\sqrt{1.542}\right)\left(\sqrt{5.516}\right)} \\
&= \frac{7.0005}{10.0269} = 0.6981
\end{aligned}
$$

$$\hat{Y}_{t+\tau} = 0.6981\hat{y}_{1,t+\tau} + 0.3019\hat{y}_{2,t+\tau}$$

$$
\begin{aligned}
Min\,Var\left[e_{t+\tau}^c\right] &= \frac{\sigma_1^2\sigma_2^2\left(1 - \rho^2\right)}{\sigma_1^2 + \sigma_2^2 - 2\rho\sigma_1\sigma_2} \\
&= \frac{(1.542)(5.516)\left(1 - 0.509^2\right)}{1.542 + 5.516 - 2(-0.509)\left(\sqrt{1.542}\right)\left(\sqrt{5.516}\right)} \\
&= 0.6285
\end{aligned}
$$

**7.17**

$$
\begin{aligned}
Min\,Var\left[e_{t+\tau}^c\right] &= \frac{\sigma_1^2\sigma_2^2\left(1 - \rho^2\right)}{\sigma_1^2 + \sigma_2^2 - 2\rho\sigma_1\sigma_2} \\
&= \frac{\sigma_1^2\sigma_2^2\left(1 - 0\right)}{\sigma_1^2 + \sigma_2^2 - 2(0)\sigma_1\sigma_2} \\
&= \frac{\sigma_1^2\sigma_2^2}{\sigma_1^2 + \sigma_2^2}
\end{aligned}
$$